BEING WITH: ESSAYS IN POETICS, ECOLOGY, AND THE SENSES

JOHN CHARLES RYAN

BEING WITH: ESSAYS IN POETICS, ECOLOGY, AND THE SENSES

JOHN CHARLES RYAN

First published in 2014 in Champaign, Illinois, USA
by Common Ground Publishing LLC
as part of the On Sustainability book series

Copyright © John Charles Ryan 2014

All rights reserved. Apart from fair dealing for the purposes of study, research, criticism or review as permitted under the applicable copyright legislation, no part of this book may be reproduced by any process without written permission from the publisher.

Library of Congress Cataloging-in-Publication Data

Ryan, John Charles, author.
  Being with : essays in poetics, ecology, and the senses / John Charles Ryan.
    pages cm
  Includes bibliographical references and index.
  ISBN 978-1-61229-491-9 (pbk : alk. paper) -- ISBN 978-1-61229-492-6 (pdf : alk. paper)
  1. Botany in literature. 2. Ecology in literature. 3. Senses and sensation in literature. 4. Literature and science. 5. Poetics. 6. Science. 7. Botany. 8. Ecology. I. Title.

PN56.B73R93 2014
580--dc23

2014014733

Cover image Credit: Nalda Searles, 'Mandala Basket' (2003)
(native plant twigs, balga bracts, balga seed heads, cloth fragments, linen threads)
Contact: naldasearles1@bigpond.com

Table of Contents

Preface: On Being With xiii

Chapter 1: The Things of Nature: Toward a Phen(omen)ology of the Seasons 1
    Introduction 1
    Revising the Australian Seasons 1
    Seasons of Things: A Phenomenology of Dwelling with/in 3
    Seasons of Our Inheritance: The Appearance of the Gregorian Model 6
    Seasons of the Southwest: The Endemic Calendar of the Nyoongar 11
    Seasons of Our Dwelling: The Indigenous Weather Knowledge Project (IWKP) 15
    Conclusion: Living with Seasonal Plurality in Australia 18

Chapter 2: Stories of Snow and Fire: The Value of a Pluralistic Environmental Aesthetic 19
    Introduction 19
    Stories of Snow and Fire 20
    Science as Narrative: Carlson's Natural Environmental Model 23
    Returning to Denali: Indigenous Cosmology as an Appropriate Narrative for Aesthetic Appreciation of Nature 25
    As with Ice, So with Fire: Science and Stories of Burning in Aboriginal Australia 28
    The Role of Journalistic Narratives in Popular Appreciation of Fire and Ice .. 30
    Returning to the Aesthetics of Burning: Examples from North America and Africa 32
    Narratives of Fire and Ice in Dialogue: Representing the Natural World in Critically Pluralistic Terms 35
    Conclusion: Toward Critical Pluralism through Narratives 37

Chapter 3: Plant Narratives and the Senses: Thoreau's Approach to the Botanical 39
    Introduction 39
    Thoreau's Multiple Narratives of Nature 40
    The Narrative Streams of Zamia Palm 42
    Nyoongar Conceptualizations of Plants 44
    Conclusion: Botanical Narratives and the Senses 48

Chapter 4: Cultures of Flora: Conserving Perth's Botanical Heritage through a Digital Repository 49
    Introduction 49
    Background to the Project 49
    Redefining, Documenting, and Conserving Urban Plant-based Cultural Heritage 51
    Bridging Tangible and Intangible Plant-based Cultural Heritage 52
    FloraCultures Methodology: Combining Archival and Digital Techniques ... 54
    Gift of Blood: Catspaws and Kangaroo Paws (*Anigozanthos* spp.) 56

Conclusion: The Practice of Botanical Heritage 58

Chapter 5: Reading Botanical Aesthetics: Embodied Perceptions of Perth's Flora, 1829 to 1929 60
Introduction 60
Sustainability and Being with Plants 61
An Embodied Aesthetics of Plants: Sensation and the Allure of the "Lower" Senses 62
Interpreting Botanical Heritage through Floraesthesis: Some Examples from Perth 65
Conclusion: Toward Aesthesis 68

Chapter 6: "The Name Blossomed": Landscapes, Habitats, and Botanical Poetry 70
Introduction 70
Landscape Poetry and Habitat Poetry 71
Reading Lansdown, Choate, and Kinsella as Habitat Poets 73
Domestic Spheres and Natural Habitats 83
Conclusion: Habitat Poetics 86

Chapter 7: Four Poems on Mushrooms: A Poetic Mycology of the Senses 88
Introduction 88
Mycotal Otherness in Science and Ecocriticism 88
Theorizing a Poetic Mycology of the Senses 90
The Elf of Plants: Emily Dickinson's "Mushroom" 92
Earless and Eyeless: Sylvia Plath's "Mushrooms" 94
Flocks of Glitterers: Mary Oliver's "Mushrooms" 96
Everywhere They Touched Us: Caroline Caddy's "Mushrooms" 98
Conclusion: The Poetry of the Forgotten Kingdom 99

Chapter 8: Plants, Processes, Places: Sensory Intimacy and Poetic Enquiry 101
Introduction 101
The Praxis of Being With 101
Interlude I: Lesueur National Park, Jurien, Western Australia 102
Understanding Parrot Bush 103
The Phenomenology of Being With 104
Interlude II: Jarrahdale, WA 105
Inside a Jarrah Tree, A Black Tunnel Reaching Skyward 106
The Processes of Being With 107
Interlude III: Anstey-Keane Damplands, WA 108
First Kangaroo Paws 109
The Poetics of Being With 109
Interlude IV: John Forrest National Park near Perth, WA 111
Sunday Zamia Swagger 112
Conclusion: The Places of Being With 112

Chapter 9: Darwin's Speculative Method: A Poetry of Science, or a Science of Poetry? 114
Introduction 114
The Speculative Method of Erasmus Darwin 114

Versified Science: Blending Poetic Imagination and Scientific Theory 117
Evolution and Darwin's Theory of Organic Happiness 122
Sneaking in Scientific Exposition: The Footnotes Scheme......................... 125
Conclusion: Speculating about Darwin's Versification of Science 126

References **128**

Index.. **144**

Acknowledgements

I acknowledge the Nyoongar people of the Southwest of Western Australia, especially Noel Nannup, Sharon Gregory, and other teachers who have generously taught me and many others about the Aboriginal Australian beliefs and practices of this part of the country.

I thank the interdisciplinary CREATEC research group at Edith Cowan University, Western Australia, for an encouraging environment and financial sustenance during the writing of these chapters and the preparation of this book. In particular, I have benefited greatly from the insights of ECU colleagues Rod Giblett, Debbie Rodan, Nandi Chinna, and Glen Phillips. The most indispensable support has come as a Postdoctoral Research Fellowship in communications and arts (2012–2015) for which I am extremely grateful to ECU. I similarly wish to thank Lancaster University, UK, where I completed my MA degree in values and the environment under the supervision of Isis Brook and Emily Brady. Their expertise in philosophical aesthetics stimulated an interest that stays with me still. My appreciation also goes to Warwick Mules for inspiring the concept of *being with* through his work on Heidegger in *With Nature* (2014, Intellect Press).

I also wish to recognize the extraordinary efforts of the editors, referees, and proofreaders of the peer-reviewed journals in which versions of the nine chapters in *Being With: Poetics, Ecology, and the Senses* have appeared, both online and in print. These journals are *Environment, Space, Place* (Gary Backhaus, David Macauley, and Luke Fischer), *Humanities* (MDPI), *Refereed Proceedings of the Australian and New Zealand Communication Association Conference* (Terence Lee, Renae Desai, and Kathryn Trees), *Australasian Journal of Ecocriticism and Cultural Ecology* (CA Cranston and Tom Bristow), *PAN: Philosophy Activism Nature* (Freya Mathews, Kate Rigby, and Alison Pouliot), *Axon* (Paul Hetherington and Jen Webb), and the *International Journal of the Literary Humanities*. Their scholarly generosity has greatly improved this collection of essays. Any error of content, style, or argument in *Being With*, however, are entirely mine.

Additional thanks go to the Royal Western Australian Historical Society for helping me locate the journals of early WA botanists and for permission to reprint Image 1. And to the Monastery Guesthouse at the New Norcia Benedictine Community for providing a wonderful, meditative space for a writing retreat during which I completed this book.

Last but not least, I am indebted and semi-permeably linked in one way or another to the places that inform this research—most notably the Southwest of Western Australia, where I have lived since 2008, and the regions of North America described in Chapter 2. To the poets, thinkers, and activists of these places, my sincere admiration. May the human (superhuman, unhuman, extrahuman, more-than-human, other-than-human) voices here and there continue to resound, thrive, and inspire.

Figures

Fig. 1 Map of Western Australia, 1886 xii
Fig. 2 Denali Panorama 21
Fig. 3 Aerial View of Kings Park and Botanic Garden 50
Fig. 4 FloraCultures Holding Page 51
Fig. 5 Kangaroo Paw Sculpture 57
Fig. 6 Balga Flower Detail 75
Fig. 7 Mudja Flower Detail 77
Fig. 8 Field of Everlastings 81
Fig. 9 Bottlebrush in Flower 81
Fig. 10 "Flora at Play with Cupid" from *The Botanic Garden* 115
Fig. 11 "Erasmus Darwin" by Joseph Wright of Derby 116
Fig. 12 *Vallisneria spiralis* from *The Botanic Garden* 119

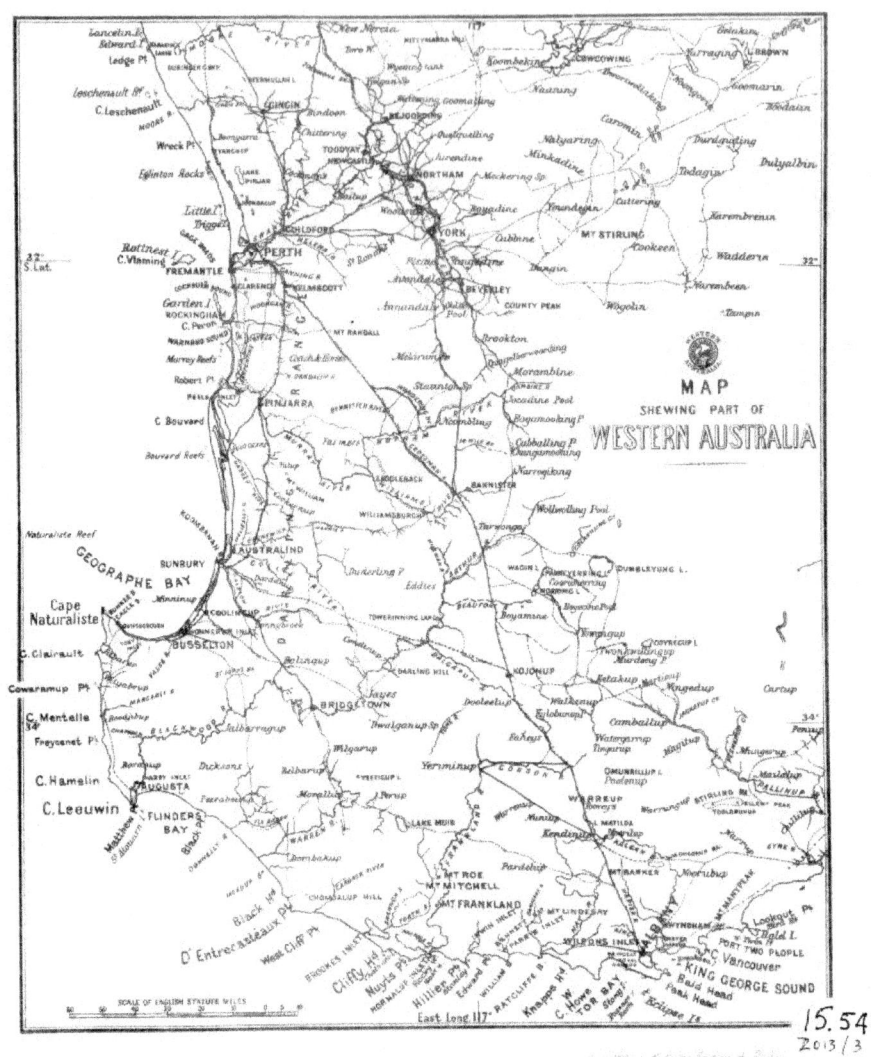

Figure 1. 1886 map of Western Australia, by E. Spiller, Government Printer, Adelaide, SA, with marginalia.
Source: Royal Western Australian Historical Society.

Preface: On Being With

I have been watching over a particular tree now for about five years. The fire-flowered West Australian Christmas Tree (*Nuytsia floribunda*) lives in a small bushland reserve in a Perth suburb, in close connection to a community of banksias and balgas typical of the Swan River coastal plain. In fact, the tree is a hemi-parasite and has to gain some of its nutrients from the roots of host plants to survive. After the long-drawn-out spring rains, the blossoms were notably vivid this year but, unlike other spring seasons, without the sweet, acrid, and stimulating fragrance I have come to associate with its kind. Each of my visits to this tree, no matter what time of year and no matter how long or short, reveals something new to me: a procession of insects harvesting its nectar, a glint of the sun on its irregular canopy, the smell of its leaves after a heavy downpour, and the tactile memory of touching its rough bark. I am increasingly getting to know the Christmas Tree through the uncomplicated yet profound choice of being with *this* individual.

The following essay collection, written over the last ten years, develops a philosophy of *being with* in relation to the botanical and mycological worlds—trees, shrubs, flowers, herbs, and mushrooms. What does it mean to *be with* (and conversely, *without*) the many beings of the world? What are the possibilities of *being with* for environmental thinking, writing, and doing? In the nine chapters comprising the book, I explore *being with* as a poetics of place (including other species) that builds on Martin Heidegger's concept of *dwelling* and related ideas developed in his collection *Poetry, Language, Thought*. A poetics of *being with* can inspire new ways of knowing and living "with nature" (Mules 2014) not beholden to ontologies of incessant speed, modernist hyper-rationality, or technocratic progress. *Being with* is a "slow" nature poetics (akin to other slow movements of today, including slow foods) involving the nuanced interaction of ecology, sense, imagination, and creativity. As a relational poetics of duration, *being with* refuses the dichotomous structures of Western thinking (human and non-human, nature and culture, country and city, body and mind), while also recognizing, in Heidegger's terms, the *dif-ference* (between us and the world) that should underlie our ethical regard for animals, plants, fungi, insects, stones, and the places they (and we) comprise, together.

As many of the chapters demonstrate, a close reading of poetic language (in the form of poetry, novels, diaries, journalistic commentaries) forms a major part of *Being With*. For, as philosopher Eleanor Helms (2008, 36) argues, language "is our *poiesis*—our authentic *making*—which allows us both to dwell in the world and keep watch over it." In a scholarly context, poetics concerns the manner in which poetry and other works impact a reader through literary techniques or contexts. Rather than purely about the interpretation of a work, poetics considers the effects of language on the reader or community of readers. While language mediates our everyday experience and is, therefore, inescapable, a poetics of the world need not be exclusively about poetry (verse, sonnet, villanelle, haiku, metaphor, simile) or other forms of language per se. Indeed, I suggest that one can (and should) have a poetics of the world without ever having read a poem (if that were possible). Simply put, in *Being With*, I propose a poetics of everyday life in which we "make sense" of the world, forging for ourselves the connections between disparate "elements" (i.e. plants and fungi) and their influence on us. This amounts to poetics as pattern-thinking. My interpretation of poetics, therefore, is *ecological* (about interrelationships), *phenomenological* (about experience and perception), and *reflexive* (the latter informs the former, and vice versa, continuously and reciprocally).

Many of us could speak at length about our personal ethics: honesty, recycling, modesty, solar energy. Moreover, the field of environmental ethics has been firmly established in the West at least since American conservationist Aldo Leopold wrote in the 1940s of "a land ethic [that] changes the role of *Homo sapiens* from conqueror of the land-community to plain member and citizen of it, [an ethic that] implies respect for his fellow-members, and also respect for the community as such." But what of our poetics? How do we sense (and make sense of) the world around us and within us? What patterns our thinking about nature? And why should we care about nature in the first place? What really motivates human behavior and environmental ethics? Self-preservation? Collective social guilt? Loss of beauty? Factual understanding of the stakes?

In this book, I suggest that the human tendency to cherish and to know more intimately—values which can be cultivated through education and experience—can be usefully thought through via the context of *being with*. Frightfully, the alternative to *being with* is *being without*: flora, fauna, fungi, places, ecosystems, water, air, seasons, and the cultural heritage emerging from biodiversity. The implications of *being without* should be obvious: species loss, extreme urbanization, habitat destruction, anthropogenic diseases, sensorially homogenized environments, the uglification of the world, the commodification of nature, the loss of heritage. At its root, *being with* recognizes the potential for human life to be naturally full (of sounds, smells, tastes, textures, and sights) in contrast to the possibility of even more profound impoverishment through hypermodern ruination of the *our* life-worlds—those we share with other-than-humans. Being with lets nature be (Mules, 2014).

Being With outlines the complex relationships and constructive possibilities between poetics, ecology, and the senses. Striking a dialogue between cultural theory, literary analysis, and environmental philosophy, I argue that an ethics of nonhuman beings, in part, necessitates a "slow" poetics of nature attuned to long-term patterns and ecological changes. Often considered passive objects, immobile things, or mere aesthetic features of the landscape, plants and fungi are integral to the life support systems of the earth and to human cultures worldwide.

Nevertheless, many members of these biological kingdoms have been reviled and, subsequently, eradicated by human societies. Modern conservation science continues to reveal that the smallest and most unassuming organisms (i.e. often the "slowest")—once treated as functionally worthless or morally deplorable—underpin the integrity and long-term viability of the ecosystems on which human life depends. Through the optimistic construction of poetics put forward in *Being With*—one which bridges science and art, and one which propounds an ethical poetics of everyday values, perceptions, and work—plants and fungi can be encountered as "companion species," to borrow Donna Haraway's term, explained in Chapter 5.

Many of the concepts in *Being With* have been developed in relation to the biota of the South-West corner of Western Australia—a region that I have been writing about since 2008. This essay collection draws partly from my ongoing research into the cultural significance of South-West plant species, published previously in the books *Green Sense* (2012), *Two With Nature* (2012), and *Unbraided Lines* (2013). The biodiversity of the South-West is of international significance. In fact, as botanical science tells us, the region is one of the most floristically diverse Mediterranean ecosystems in the world. In terms of its physical scope, the South-West extends from Shark Bay (in the upper northwest corner of the State) to Israelite Bay east of Esperance (in the southeast corner), and includes Geraldton, Bunbury, Augusta, Albany, and the metropolitan area of Perth. In the late nineteenth century, botanist Baron von Mueller deemed the South-West a "botanical province" because of its distinctive floristic communities and high rates of plant endemism. Now showcased as a global biodiversity "hotspot," the South-West exhibits a remarkable degree of faunal and floral uniqueness—for example, plants such as Mangles kangaroo paw (*Anigozanthos manglesii*) that occur nowhere else on the planet.

Moreover, on a technical level, the South-West is the only globally recognized Australian hotspot. This is because nearly 40 percent of its plants are "endemic"—found to occur in uncultivated conditions only within the geographical delimitations of the region. Another compelling fact is that the South-West province accounts for 80 percent of the endemic plants of Western Australian, including iconic flora such as the underground orchid (*Rhizanthella gardneri*), poverty bush (*Eremophila alternifolia*), and the WA Christmas tree (*Nuytsia floribunda*), just described. The genesis of the region's botanical diversity can basically be attributed to its nutrient-poor soils and intensely dry, hot summers. The South-West is a case study in environmental adaptation. Conservation science aside, the biodiversity and endemism of the region has also shaped the works of poets, artists, illustrators, diarists, journalists, tourists, pastoralists, and other commentators of the colonial and post-colonial periods. As such, plants and fungi can be viewed as creative agents of meaning-making and cultural production—as agentic.

Being With comprises nine chapters and is woven together with a transdisciplinary variety of themes, including aesthetics, imagination, intimacy, seasonality, heritage, habitats, science, botany, and mycology. Building on the philosophies of Heidegger and Merleau-Ponty, Chapter 1 proffers a phenomenology of the seasons through a case study of the six traditional Nyoongar seasons of the South-West and the Indigenous Weather Knowledge

Project (IWKP). This chapter argues that being with the endemic seasons of a place involves attending to specific changes over time in relation to ecological processes and cultural understandings. Turning toward examples from outside Australia, Chapter 2 suggests that the narratives we adopt within (or exclude from) our poetics influence the environmental values we go on to develop. A critically pluralistic narrative framework encompasses Indigenous, scientific, folkloric, and popular stories, in this chapter, of snow and ice, leading to a more diversified foundation for aesthetic appreciation. Furthering the narrative theme, Chapter 3 pursues the theorization of narrative multiplicity through the botanical writings of Henry David Thoreau and the relevance of his inclusive yet critical model for the appreciation of South-West plants. A sensorial, embodied aesthetics of plants is a crucial mode of being with the botanical world.

An applied poetics of *being with* is the preoccupation of Chapters 4 and 5 in which I discuss a project named FloraCultures in relation to themes of heritage, memory, aesthetics, sense, and plants. Here, the work of archiving is underpinned by the ethos of *being with*. Conserving the plant-based cultural heritage of a place presents another dimension of our relationship to living plants in biodiverse places such as the South-West. Turning from heritage to poetry, Chapter 6 looks at a cross-section of South-West botanical poetry through a close textual reading of works by Andrew Lansdown, Alec Choate, John Kinsella, Dorothy Hewett, Glen Phillips, and Tracy Ryan. As a fusion of sense, science, and ecology, a habitat poetics inherently involves sense engagement with plant life, sustained observation over time, and a willingness to be with more-than-humans. Continuing the focus on ecopoetry, Chapter 7 proposes a *poetic mycology of the senses* through the lens of multispecies theory informed by Donna Haraway, Anna Tsing, and others. *Being with* the fungal world requires sustained embodied experience combined with acute ecological knowledge. Shifting toward my own poetic work, Chapter 8 takes the form of a practice-led rumination of concepts of process, phenomenology, poetics, and place. In this chapter, I offer a creative arts model for putting *being with* in practice. Concluding the collection, Chapter 9 highlights the entanglement of imagination, poetry, and botany through a reading of Erasmus Darwin's *The Botanic Garden* and *The Temple of Nature*. Clearly moving out of the South-West Australian context, this chapter points to the pivotal role of imagination in a poetics of *being with*.

John Ryan
New Norcia, Western Australia

Chapter 1: The Things of Nature: Toward a Phen(omen)ology of the Seasons

Introduction

Since European settlement in Australia, the Western calendar has poorly accounted for the seasonal nuances and multiple temporalities of the land. Beginning with Tim Entwistle's recent proposal to revise the four-season Australian norm, this chapter traces the emergence of the Western calendar in Europe and its institutionalization "Down Under." With its emphasis on land-based calendars, the Indigenous Weather Knowledge Project (IWKP) is a partnership between Aboriginal communities and the Bureau of Meteorology aimed at preserving and promoting knowledge of the endemic seasons of Australian regions. As the most recent addition to the IWKP, the six-season Nyoongar calendar of the South-West of Western Australia is based on meteorological conditions (ecological time), such as wind directions and temperatures, but also on the procurement of food, maintenance of cultural knowledge, and performance of ceremonies (structural time). Through the fusion of phenomenological (experiential, sensory, place-based, actual) and phenological (cognitive, visual, enumerative, digital) approaches, the endemic seasons of Australia can be appreciated in their depth and extent.

Revising the Australian Seasons

In a recent article in *Australian Geographic*, Tim Entwistle, Director of Conservation at Kew Gardens and former Director of the Sydney Botanic Gardens Trust, proposes a five-season model for Australia. Entwistle's schema includes a weightier four-month summer (December–March), a slenderer two-month autumn (April–May), and a compressed two-month winter (June–July). Revising and reassigning the antipodean seasons, he divides spring into a two-month "sprinter" (August–September) and two-month "sprummer" (October–November). Entwistle's revisionist five-season thinking unmistakably emphasizes the Australian summer, comprising one-third of the calendar year in his schema. Additionally, spring (reconceptualized through the neologism "sprinter") begins in August—one month earlier than its four-season counterpart—to correspond to the flowering of native plants in many parts of the country. Critical of the European temporal grid, Entwistle regards seasons as "cultural constructs reminding us that there are cyclic changes in the environment" (quoted in Duncan 2011, April 1). Judging from his revisionist proposal, the usual constructs— spring, summer, autumn, winter—are unsatisfactory when invoked "Down Under."

In Entwistle's view, the Australian seasons require reconsideration, optimistically leading to new modes of seasonal awareness. On the surface, five seasons more sensibly accommodate the natural cycles of the Australian biorhythm. His ecologically inspired calendar, in part, adjusts its demarcations to the chief flowering time of Australian native flora as a whole. However, while I recognize that Entwistle's five-season tender is praiseworthy, any template for generalizing the Australian seasons inevitably becomes ensnared in the mode of cultural construction that it seeks to overcome. In its reconfiguring and compartmentalizing of the cyclical progression of time, Entwistle's model reproduces the ineluctable weaknesses of a single seasonal paradigm for a land mass as vast and diverse as Australia. The cultural construction of the seasons— exemplified by the Gregorian or Christian calendar now used by nearly all Western countries (Aveni 1990, 116-117)—implies a singular and monologic rendering of seasonality, largely dislocated from the ecological nuances of regions, bioregions, places and sites.

Whether four or five in number, an Australian seasonal standard needs to be thoughtfully and continually counterbalanced by local knowledge of the seasons, encapsulated within Indigenous ecological calendars. While an incomplete formulation of Australian seasonal plurality, the five-season model's encompassing of regional land-based calendars offers a promising way forward and a basis for deeper understanding of the seasons. In short, broadly based models of seasonality—including Entwistle's—can be enhanced through sustained reference to the tacit embodied knowledge encoded within indigenous calendars. Hence, in response to Entwistle, a dialogic perspective on the seasons considers multiple places, scales, temporalities, ecologies, bodies, and cultural traditions. As a counter-example to the five-season proposal, the Indigenous Weather Knowledge Project (IWKP) offers a means for counterpoising any single, fixed, Australia-wide system. The project aims to consolidate and preserve the seasonal knowledge of Aboriginal cultures in consultation with their elders (Australian Bureau of Meteorology 2010). One of the practical outcomes of the IWKP is the digital documentation of indigenous calendars on the project's website. Accordingly, the IWKP is poised to educate the public through open-access information about land-based or endemic seasons.

The purpose of this chapter is to trace the back-story to Entwistle's call to reformulate the Australian seasons. In sketching the context broadly, I begin with the origin of the Gregorian construct, alluding to its importation to Australia as part of the processes of colonization since the 18th century. Here, I argue that the singular model of the four seasons displaced (and potentially still displaces where traditional knowledge networks are threatened or otherwise compromised) the multiple modes of season-reckoning in Australia. I then go on to consider the twin notions of endemic seasonality and indigenous calendars through historical reflection on the six-season Nyoongar calendar (Bates 1985, Bindon and Walley 1992, Moore 1884/1978, Ryan 2012b). The Nyoongar are the Aboriginal people of the South-West corner of Western Australia (Green 1984, South West Aboriginal Land & Sea Council 2009, Van den Berg 2002). After the case study of the Nyoongar calendar and its embodied, phenomenological dimensions, I conclude with a brief analysis of the IWKP and the larger context out of which it emerged.

Throughout my longitudinal discussion of the Australian seasons—from long-standing Indigenous traditions, to the Gregorian import, and to contemporary modes of Australian season-telling, represented by the IWKP—I propose and develop the portmanteau "phen(omen)ology" in relation to the seasons. I argue that the IWKP is best conceived of as an online *phenological* template that gives actual human *phenomenological* exploration of the seasons a reference point for contemporary Australians interested in acquainting themselves with the endemic seasonalities of their places. Both phenology and phenomenology are entwined processes that are essential to grasping the meaning of endemic seasonality in Australia and to learning to live with the seasons more consciously and concertedly.

Seasons of Things: A Phenomenology of Dwelling with/in

Before addressing the back-story to Entwistle's five-season call, I will set out a philosophical position on the seasons through the phen(omen)ology framework. I ask: How should we rethink the four Australian seasons in a manner that is sensitive to Australian places and cultures? How can individuals learn about the seasonal specificities of where they live in connection to national standards of seasonality—whether four or five? And, how can settler culture in Australia—steeped in four-season perception—begin to appreciate and hopefully "dwell" with and in the endemic seasonalities of regions, as described and practiced by Aboriginal cultures? As suggested in the previous section, the incorporation of land-based seasonal knowledge into Australian culture through indigenous calendars is optimally approached phenomenologically and phenologically. The former occurs as an individual's experience of the seasons through sight, hearing, touch, taste, and olfaction: as physical sensations registered in the body sensorium. The latter refers to cognitive awareness of the progression of ecological events in time linked to changes in plants, animals, the wind, constellations, and other biotic and abiotic phenomena. To begin with, phenomenological engagement centralizes immediate physical knowledge of the endemic seasons of a place: seeing, tasting, feeling, touching, and smelling the seasons, in their tangible manifestations, as they unfold. In adumbrating a phenomenology of the seasons, Martin Heidegger's notions of dwelling (1971, 143-159) and "the thing" (1971, 163-180), in conjunction with Maurice Merleau-Ponty's embodied subjectivity (2012), are valuable frameworks. Moreover, recent theoretical developments in phenomenological geography (Bender 2002, Tilley 1994, 2010) and phenomenological approaches to literary and cultural studies through the concept of "embodied temporality" (Ryan 2013, Chapter 1) also provide key conceptual positions.

Here, it is crucial to recognize that indigenous ecological calendars—such as those of the Nyoongar and Yawaru of Western Australia—are lived (and living) calendars. The sensory cues of ecological calendars are intrinsically connected to intimate seasonal knowledge. When navigated phenomenologically in the environment, these cues—e.g., the ripening of the cocky apple and its sensory materializations through pungent smell, sweet taste, and pleasing image—signal the changing of the seasons integrated to human bodily resonances. Thus, for Australian settler society, a return to endemic seasonality entails corporeal participation in places of dwelling. My phenomenological call is heightened by the fact that ecological indicators of seasonal onset and transition vary annually according to manifold factors, such as rainfall, made even more irregular by the seasonal disruption associated with climate change (CSIRO 2011, Steffen et al. 2009, 68). To state the need differently, in order to appreciate endemic calendars, one must recognize their indications physically and immanently; a phenomenology of the seasons is therefore bodily, multi-sensory, and integrative of nature and culture.

A phenomenology of the seasons attends to the "things" of nature (animals, plants, rain, wind) which, in their sensuous being, announce the seasons and their passage. Heidegger's "dwelling" is a key concept, developed in his essay "Building Dwelling Thinking" (1971, 143-159). Through human place-dwelling, the presencing of the seasons comes forth and registers sensorially. For Heidegger, dwelling is the necessary quality of being. In examining the notion of dwelling in relation to Heidegger's articulation of "the thing" (1971, 163-180), a philosophy of the seasons situates the vital things of nature—in their particular modes of being as sensorially manifested—before the fixed, mathematical, and political logos of the Gregorian model. Heidegger argues that to dwell means "to remain, to stay in a place" (1971, 144). "To dwell" implies the verb "to be" and "the way in which you are and I am, the manner in which we humans *are* on the earth [italics in original]" (1971, 145). To this effect, Heidegger links etymologically the Old English and High German word *bauen*—for building—to "dwelling" and, more compellingly, to "be" such that "I am" signifies intrinsically "I dwell." More apposite to the vitality of seasonal being in place, *bauen* connotes "to cherish and protect, to preserve and care for, specifically to till the soil, to cultivate the vine" (1971, 145).

As integrated beingness, dwelling consists of the fourfold oneness of earth, sky, divinities, and mortals; each implies the other so that, for example, thinking of earth entails thinking of sky and divinities. For Heidegger, 'earth' refers to "blossoming and fruiting," whereas 'sky' connotes "the course of the changing moon…the year's seasons and their changes…the clemency and inclemency of the weather" (1971, 147). To dwell phenomenologically in the seasons is to leave "to the seasons their blessing and their inclemency" (1971, 148)—to apprehend the seasons without exerting predetermination, control, or constraint; to allow the seasons to "presence," in their originary places to the human sensorium in the act of season-telling. Moreover, dwelling is "always a staying with things" (1971, 149). Heidegger points to the exigency of dwelling in the early twentieth century in which humanity "*must ever learn to dwell* [italics in original]" (1971, 159). In developing a "phenomenology of landscape," Tilley observes that, for Heidegger, "spaces open up by virtue of the *dwelling* of humanity or the *staying with things* that cannot be separated: the earth, the sky and the constellations, the divinities, birth and death [italics in original]" (1994, 13). Additionally, Tilley notes the "total social fact of dwelling, serving to link place, praxis, cosmology and nurture" (1994, 13) in Heidegger's thinking. The primacy of Heideggerian dwelling, in Tilley's analysis, centralizes the human body as the plenum of apprehension within the landscape and, by extension, within the seasons. Dwelling *with* and *in* the seasons is a habitus of being that reflects the entangling of ontology, cosmology, plants, animals, insects, and human consciousness.

What does Heidegger mean by "things"—a word which, in common parlance, tends to denote the inanimate stuff or commoditized objects of the world rather than the living beings calling forth the seasons in their sensuous natures? In the essay "The Thing," Heidegger differentiates between objects and things. An object is "that which stands before, over against, opposite us" (1971, 166) as the objectified "standing reserve" of technological enframement or *Gestell* (Heidegger 1977). In comparison to the instrumentally derived value of objects, a thing "stands forth" (1971, 166) agentically in its own right, manifesting the fourfold oneness of earth, sky, divinities, and mortals in Heidegger's philosophy. "Thing" refers to the presencing of an essential nature of living and non-living entities (1971, 172). As the gathering of oneness, the thing also traces the process of bringing forth Heidegger's notion of fourfold unity: "The thing stays—gathers and unites—the fourfold" (1971, 178). While they can be dead matter, things can also be animate, in Heidegger's view, as "things, each thinging from time to time in its own way" (1971, 180). Hence, rethinking the Australian seasons means to dwell with things through the seasons in the places that circumscribe each: the cocky apples and the wild yams in Yawuru country north of Broome, Western Australia, or the banksia and red gums in Nyoongar country near Perth, for instance. The "thinging" of seasonal things is their presencing through their sensory manifestations—their ripening, effusions, stridulations—at particular times of the year. The human body, thus, is a sensing agent of the seasons in conjunction with knowledge of phenological details, such as those recorded by the IWKP, including such details as flowering, fruiting, nesting, and molting times, for example.

The concept of the human body as the plenum of sensory apprehension, while weak in Heidegger's account of the presencing of things, is more fully articulated in Merleau-Ponty's work, particularly *Phenomenology of Perception*. Part One, "The Body" (2012, 95-205), outlines Merleau-Ponty's corporeal phenomenology—a complex philosophical position drawing from psychology, which I will only describe briefly here in order to suggest a complementary conceptual perspective to "the thing." In comparable terms to Heidegger, Merleau-Ponty (2012, 61) avers that "sense experience is that vital communication with the world which makes it present as a familiar setting of our life." Sense experience signifies the presencing of things. Moreover, embodiment—living in one's senses and knowing/navigating the world sensuously through one's body—is a condition of "the temporal structure of being in the world" (2012, 86). Time is integral to the twin conditions of embodiment and being. Importantly, Merleau-Ponty's account of phenomenology attends to human sensation. As part of the plenum of apprehension, kinesthetic sensations result from the movements of one's body in space (Merleau-Ponty 2012, 96). On the whole, Merleau-Ponty's concern is for the incarnate subject; his phenomenology counters the objectification—i.e., dissection, commoditization, marginalization—of the living body (Glendinning 2007, 134). Instead, the human body, rather than an object in the world, is the primary means through which we communicate with others and our environments (Glendinning 2007, 135).

Extending Heidegger and Merleau-Ponty, recent work in phenomenological geography and embodied cultural studies provides an additional conceptual foundation for a phenomenology of the seasons. Barbara Bender outlines a perspective on geographical research "where the time duration is measured in terms of human embodied experience of place and movement, of memory and expectation" (Bender 2002, 103). Bender implies that, in lieu of fixed points of reference for season-keeping, the human body acts as an ever-open sensorium, marking the seasons somatically in their fugue-like progression over time. I have previously termed this condition of being in the world as "embodied temporality" or the "sense for time and seasons engendered through physical, multisensorial interactions with place" (2013, p. 8). In developing the concept of embodied temporality, I drew from Australian ethnobotanist Philip Clarke's work on Aboriginal "calendar plants" to describe seasonal things that provide—often simultaneously—a time-keeping measure and a source of physical sustenance. Similarly, Christopher Tilley argues that human embodiment—entailing multi-sensory openness to the things of the seasons—is essential to a phenomenology of place: "A phenomenologist's experience of landscape is one that takes place through the medium of his or her sensing and sensed carnal body" (Tilley 2010, 25), a characteristically Merleau-Pontian position. Extending Tilley's framework, a phenomenological approach to the seasons implies a "dialogic relationship between person and landscape" which stresses the materiality of landscapes as "real and physical rather than simply cognised or imagined" (2010, 26). In Heideggerian terms, the materiality of earth is the "blossoming and fruiting"—the ecological processes which underlie the presencing of things. For Tilley, a number of attributes and dispositions define phenomenological being in landscape, including "perception (seeing, hearing, touching), bodily actions and movements, and intentionality, emotion and awareness residing in systems of belief and decision-making, remembrance and evaluation" (1994, 12). All of these modes of experience and cognition play out in a phenomenology of the seasons.

Seasons of Our Inheritance: The Appearance of the Gregorian Model

Turning from phenomenology for a moment, this section outlines the emergence of the twelve-month, four-season Gregorian calendar (also known as the Christian or Western calendar) from the Julian calendar of the ancient Romans. Why should Entwistle go through the trouble of redefining the Australian seasons? What's wrong with the four season score—the venerable subject of much European and North American cultural reverie—in Australia? The aim of this section is to follow Entwistle's proposal back to the origin of the four seasons and to argue that phenomenological, place-based awareness is not vital to the Western calendar that most of us use on a daily basis. In fact, the global transition to the Gregorian calendar took until the early 1900s to reach completion. In 1582, the transition was instigated when the Gregorian calendar ("new style" (*N.S.*)), replaced the Julian calendar ("old style" (*O.S.*)) (Hawkins 1751). This erasure of an "extra" ten days—produced over time by the Julian system—corrected cumulative calendrical "shifts since Caesar" (Feeney 2007, 150). The Gregorian calendar is now the international civil calendar and derives from the 16th century European desire to normalize Catholic and Protestant ceremonial dates (Doggett 1992, 580). In the Julian and Gregorian schemes, the four seasons—each approximately three months in duration—correspond to two equinoxes and two solstices per annum. Whereas land-based calendars must be experienced phenomenologically to be appreciated and often have fewer or greater than four seasons, the Gregorian model largely stems from structural, religious, political, and, later, colonial prerogatives.

The current use of the Gregorian calendar and associated four seasons in Australia can be traced to the British adoption of the calendar in 1752. Mathematically moderated, the Gregorian seasons are based on the solstices and equinoxes. Winter solstice is the shortest day, while summer solstice is the longest; the two equinoxes occur when night and day are of equal length. The Gregorian calendar—which is the underlying template for the four Western seasons—constitutes a grid-like temporal imposition on the seasonally diverse places comprising the Australian land mass. The institutionalization of the calendar is an aspect of the colonization of time—which belies the mismatch, at the core of Entwistle's call, between the diverse climates of Australian regions and the four-season overlay.

The Gregorian calendar and its Julian precedent are structural devices for reckoning time. Anthony Aveni (1990) discerns between structural and ecological time in order to identify different modes of season-reckoning, as well as the colonizing intersection of Western and indigenous calendrical systems. Aveni (1990, 174) defines ecological time as "temporal knowledge…determined by the individual as a participant in organized society" which encompasses "events in the natural world that portend change" (176). Cyclical and integrative of culture and nature, eco-time foregrounds occurrences in the natural world: "The time marker—whether flood, worm, or stars—is recognized to have a seasonal cyclic rhythm independent of human action" (Aveni 1990, 176). Whereas eco-time relates "the response of human behavior to the cycles of nature" (Aveni 1990, 177), structural time prioritizes the rituals and behaviors that regulate societies (181). In other words, structural time is based on socially significant reference points—for instance, rituals and ceremonies. For Aveni, indigenous calendars tend to coalesce ecological and structural time-keeping, leading to nuanced modes of season-reckoning that are subjective, perceptual, fluid, and potentially variable from year to year.

The meaning and function of a calendar are linked to establishing temporal predictability by controlling time, if that is even possible. Agnes Michels (1967, 9) defines a calendar as "a device for measuring time, by which [people] can plan for the future and keep a record of the past." Comparably, L.E. Doggett defines a calendar as "a system of organizing units of time for the purpose of reckoning time over extended periods…some calendars are codified in written laws [i.e. the Gregorian]; others are transmitted by oral tradition [i.e., the Nyoongar, traditionally]" (Doggett 1992, 575). Aveni (1990, 6) states that the underlying premise of a calendrical system is that a "temporal order" already exists in the natural world. A calendar merely identifies, exposes, and codifies this order. By establishing a structure for capturing and controlling the order, an institutionalized calendar avoids the problem of variation in seasonal durations in different places within a geographical area as vast as Australia. The problem of variation, according to structural thinking, is intrinsic to the subjective sensory reckoning of seasons, as evident in many Indigenous calendar systems (Aveni 1990, 6). In differentiating between structural time and ecological time, Aveni (1990, 123) emphasizes that the seasons overlap in reality; their edges are not hard and fast and do not strike firmly at certain calendrical nodes. This overlapping denotes "a sense of instability to the event sequences that make up the cycle of nature's behavior." Such instability in nature, however, for Michels (1967, 9-10), renders the (northern hemisphere) seasons an unsound basis of "only relative value" for a calendar: "although the seasons proceed in a regular sequence from year to year, they may vary considerably in length owing to variations in the weather." Moreover, to compound the difficulty of seasonal standardization and the need for a uniform system not derived from an ecological perspective, the "seasons also vary locally" (Michels 1967, 9-10)—which is certainly the land-based reality in Australia.

Four-season thinking is evident in the writings of the English Saint Bede (also known as the Venerable Bede, ca. AD 672–735). He connects the four seasons to the temperate conditions of the northern hemisphere and also to the four humors of the human body. For Bede, the seasons firstly derive from the English climate as the proper markers of the temporal order:

> The seasons [*tempora*] take their name from this temperateness; or else they are rightly called *tempora* because they turn one into the other, being tempered one to another by some qualitative likeness. For winter is cold and wet, inasmuch as the Sun is quite far off; spring, when [the Sun] comes back above the Earth, is wet and warm; summer, when it waxes very hot, is warm and dry; autumn, when it falls to the lower regions, dry and cold. (Bede 1999, 100)

Bede (1999, 100–1) then characterizes the human body a "microcosm" and "a smaller universe" in which the four humors—blood, black bile, red bile, phlegmatic humors—correspond to the four seasons. In his schema, certain humors manifest during certain seasons. The four qualities of hot, cold, wet, and dry—which couple to produce the conditions of the seasons—constitute the human humors as well. Bede associates qualities and humors with the seasons. While an embodied seasonal philosophy, Bede's thinking reiterates the quarterly division of the year implied in the ancient Roman term *tempora annu* or "times of year" (Holford-Strevens 2005, 80). Thus, Bede's humoral philosophy speaks of the provenance of the four seasons in northern hemispherical climates and bodies.

In B.C.E. 46, Julius Caesar replaced the ten-month Roman lunar calendar with a twelve-month system (Fredregill 1970, 13). Caesar's schema, which became known as the Julian calendar, averaged 365.25 days per year (Fredregill 1970, 14). As the ancient precedent for the modern calendar, it comprised twelve months, although they were denoted by somewhat different names (e.g., *Sextilis* rather than August). The main shortcoming of the Julian calendar—addressed by the Gregorian reform—was calendrical drift: the tropical year measured approximately 365.24219 mean solar days (Richards 1999, 239). Pointing to the discrepancy between Gregorian and Julian calendars, Fredregill (1970, 14) terms the Julian calendar "slow." In calculating slightly more days in the calendar year than the tropical year, the Julian system caused annual events to fall earlier in the calendar year at a rate of one day per 128 years (Richards 1999, 239). To its discredit, the average Julian annum comprised slightly too many days. Of temporal and religious concern, the actual vernal equinox began occurring in advance of its calendar date March 21, and astronomical new moons were reckoned earlier and earlier (Richards 1999, 352). Of particular concern for the medieval Church, calendrical drift caused Easter to fall on unsuitable days (Richards 1999, 249).

In A.D. February 1582, Pope Gregory XIII introduced the Gregorian calendar, instigating the Julian reformation by a bull known as *Inter Gravissimas* (Duncan 2011, April 1, 261-289, Methuen 2008, 61-73, Richards 1999, 239-256, Willes 1700). In consultation with the astronomer Ignazio Danti (1536–86), Gregory became certain that the equinoxes were falling on incorrect days as a result of Julian drift (Richards 1999, 241). By A.D 1582, the accumulated error of the Julian drift tallied more than ten days. In an edict issued eight months before the calendar reform would be instituted, Pope Gregory XIII corrected the ten-day error, mandating that October 15, 1582 revert to October 4, 1582. This reformation eliminated about ten days of Julian drift, accumulated over 1,600 years since the institution of Caesar's calendar (Duncan 1999, 261-262). Through this mandate, Gregory advanced the recommendations of the Council of Trent; although it was on the agenda of the Council, calendar reform was not sufficiently carried out until the papal decree (Richards 1999, 241).

Physician and astronomer Aluise Baldassar Lilio (1510–76) designed the Gregorian calendar for Pope Gregory (Richards 1999, 243). To correct the Julian drift, Lilio recommended that the first year of each century skip the leap year, except for years, such as 1600 and 2000, that could be divided evenly by 400 (Fredregill 1970, 14). The Gregorian reform mandated that the leap year occur every four years, but not during these particular years. It also included standards for calculating Easter, based on a revised table of new and full moons (Doggett 1992, 583, Richards 1999, 352), and assigned the beginning of spring to March 21 (Borst 1993, 103). Considering the calendar's relevance now, David Duncan (1999, 289) calls the Gregorian scheme "the world's calendar: a code for measuring time that today all but the most isolated peoples use as the global standard for measuring time." In comparable terms, E.G. Richards (1999, 256) comments that, following its introduction to Britain in 1752, "the Gregorian calendar was later taken to the four corners of the globe on the back of the British Empire. It is now all but universally used." In comparison to the Julian, the Gregorian system preserves three days every 400 years, allowing the activities of Western cultures to align almost uniformly with the solar year until A.D. 4000.

Bonnie Blackburn and Leofranc Holford-Strevens (1999, 682) summarize, in *The Oxford Companion to the Year*, that the "adjustment was necessary because the Julian year, consisting of 365 days, with a 366th day added every fourth year, has an average length of 365 days 6 hours, which is some 11 minutes 12 seconds too long, causing Julian dates to fall progressively further behind the sun." However, the Gregorian schema was not instantly adopted by all Western countries. It took approximately 300 years to become the calendrical norm and was met with social, political, and religious resistance (Donaldson 1996b, 95). In England, the reform sparked controversy, as the opposition's oft-cited motto attests: "Give us back our eleven days." A British Act of Parliament in 1752 introduced the Gregorian calendar or the "new style" (Richards 1999, 252-56). Britain's decision came 170 years after the rest of Europe, making it one of the last European countries to do so. The Act (24 Geo. II, ch. 23) was passed "for regulating the commencement of the year, and for correcting the calendar now in use" (quoted in Richards 1999, 253). Presented to Parliament by Lord Chesterfield, it became law on May 22, 1751 (Richards 1999, 253). Accordingly, 12 days were "eliminated" when September 14, 1752 reverted to September 2, 1752 (Feeney 2007, 151, 281, Duncan 1999, 277-78).

After its legalization in Britain, the Gregorian calendar was distributed to the colonies, including North America and, later, Australia. The standardization of season-reckoning in Australia culminated in the Meteorology Act of August 1906 and, subsequently, the creation of the Bureau of Meteorology in 1908 (Australian Bureau of Meteorology 2008a, 7). In 2012, the autumn equinox in Australia was March 20; the winter solstice, June 21; the spring equinox, September 23; and the summer solstice, December 21 (Australian Bureau of Meteorology 2013). However, rather than following the solstices and equinoxes in determining the start dates for seasons, Australia uses the international meteorological definition for the southern hemisphere. This mandates three-month "meteorological" or "calendrical" (rather than astronomical) seasons beginning the first of each month: September 1 (spring), December 1 (summer), March 1 (autumn), and June 1 (winter). The Australian convention makes the highly statistical process of record-keeping—as regulated by the Australian Bureau of Meteorology—more convenient and consistent.

Entwistle's initiative to rework Australian season-keeping responds to the imperialist history of the Gregorian calendar and reflects his awareness of the indigenous calendars and endemic seasons of Australia that preceded colonization. However, his criticism of the four seasons down-under is not new. In the mid-1990s, Steve Symonds, a spokesperson for the Weather Bureau of New South Wales, commented bluntly that:

> We [settler society] are cultural imperialists and we have just said what we want the weather to be. We came out here and said that there are four seasons in Europe so four seasons there should be here. Why should there be four seasons in Australia just because there are four seasons in London? (quoted in Donaldson 1996a, 204)

As this section has detailed, the Gregorian calendar—applied to the immense landmass and cultural diversity of Australia—reiterates the processes of colonization and forever inscribes a history of religious conflict and ecological repression. The Western calendar disregards the very ground of places and the materiality of things entangled with human temporal perception. In his analysis of Indigenous calendrical systems, the anthropologist Alfred Gell (1992, 313) avers that "the intertwining of calendars and power [...] extends to the processes of colonial subjugation." The importation of the four seasons to Australia—originating in the Julian drift, the Gregorian reform, and the dissemination of the calendar through British empire—made possible the erasure of the endemic seasons of Aboriginal cultures within the nationalistic paradigm of season-keeping. However, as the next section will highlight through a case study of the Nyoongar people of the South-West region of Western Australia (WA), vibrant traditions of endemic seasonality endure, despite the impact of the colonial imposition. Indigenous seasonal traditions are necessary counterpoints to any broadly applied, national seasonal paradigm—whether four or five.

Seasons of the Southwest: The Endemic Calendar of the Nyoongar

Traditions of endemic seasonality—along with the cultural integrity underlying them—should not be overshadowed by national standards—revisionist or Gregorian. Like the Western calendar, seasonal calendars or "indigenous ecological calendars" are cultural constructs—"timetables that divide the year into seasons and describe expected conditions and resource availability" (Prober, O'Connor, and Walsh 2011, 2). Yet, a land-based seasonal calendar, unlike the Western calendar, is intrinsically connected to the ground—the ecology and culture of a place, and the corporeal things of nature which announce the seasons (Usher 2000). In contrast to the four-season Western regime, "indigenous calendars," as Tim Entwistle concedes in *Australian Geographic*, more appropriately reflect regional Australian climates than the globalized four-season schema formulated in Europe. Australian indigenous calendars offer the vital complement to Entwistle's revised calendar. Aboriginal cultures have unique place-based systems of season-keeping, recognizing two, four, six, seven, and nine seasons, for example (Clarke 2007, 54-59). The danger of Entwistle's proposal is that his new model, with its relatively minor reorientation toward native plants, will simply substitute in for the Gregorian scheme—the complex nuances of each indigenous calendar again rendered one-dimensional by the imposition of a "fixed system of reference" over the entire country (Prober, O'Connor, and Walsh 2011, 2).

Derived from European political, religious, and climatic circumstances, the Gregorian calendar is an apparatus of colonization that has been misapplied in Australia and "staunchly retained" since the 1800s (Clarke 2007, 54). In contrast, the endemic calendars of Australian Aboriginal people offer pathways to ecological time—foregrounding events in the natural world—and structural time—relating the seasons to events of social significance, including ceremonies and festivals. The Nyoongar calendar of the South-West is a living system of time-keeping that signifies phenomenological engagement with the environment. Here, the presencing of the things of nature—wind, temperature, fire, flora, fauna—reflects the nuances of seasons.

This section begins with historical interpretations of the six Nyoongar seasons, recorded by Western Australian settlers and colonists, then shifts to contemporary explanations of the traditional seasons by Nyoongar elders and teachers. In previously proposing "embodied temporality" through my analysis of the Nyoongar seasons, I (2013) have interrogated a variety of historical sources, including the diaries of Albany-based doctor and settler Scott Nind (1831/1979), lawyer and farmer George Fletcher Moore (1884/1978), and early twentieth-century ethnographer and journalist Daisy Bates (1985). Building upon my initial historical research, this section will introduce material from George Fletcher Moore's diaries, as well as extracts from the published journals of the Benedictine monk Dom Rosendo Salvado and statements from colonial-era Western Australian newspaper articles referring to the Nyoongar seasons.

In Aboriginal Australia, according to Clarke (2007, 54), totemic associations, burning regimes, celestial movements, animal behaviors, wind patterns, temperature shifts, flowering phases, and rainfall levels together announce the arrival of each season. Instead of the measuring of time that is intrinsic to the Gregorian calendar, Aboriginal peoples apprehend environmental changes corporeally in order to mark the movement of the seasons (Clarke 2009, 94). Unlike the Western calendar, Australian "bush calendars" have between two and nine divisions, and the duration of each season varies annually (Clarke 2009, 95). Prior to European settlement, Nyoongar people gathered plant foods and hunted animals according to a six-season calendar, with whole camps moving into areas when particular foods became harvestable (Nannup and Deeley 2006, Rusack et al. 2011, Stasiuk and Sillifant 2005, Tilbrook 1983, 3). In Albany, oral histories describe the local Nyoongar tradition of movement with the seasons from the coast in the summer to the inland in the winter (Tilbrook 1983, 145). Traditional Nyoongar seasonal awareness "comprises organized artisanal knowledge gained through observation and adjustment over timeframes of thousands of years, often strongly linked with an ontology such as that shaped by the 'Dreaming'" (Prober, O'Connor, and Walsh 2011, 2).

Drawing from historical sources, including key records written by Bates, Nind, and Moore, Neville Green (1984, 10-11) provides a summary of the six Nyoongar seasons and their differing orthographies in Perth and Albany, Western Australia. In Perth, about 250 miles northwest of Albany, *Birok* is comparable to early summer (the first summer) and comprises December and January; in Albany, the season is known as *Meerningal*. *Burnoru* is the Nyoongar late summer (the second summer) and comprises February and March; in Albany, known as *Maungernan*. *Geran* includes the autumn months of April and May; known as *Beruc* to Albany Nyoongar people. *Maggoro* includes the winter months of June and July; known as *Meertilluc* in Albany. *Jilba* refers to the spring months of August and September; *Pourner* in Albany. Finally, *Kambarang* encompasses the spring months of October and November; denoted as *Mokkar* in Albany. The six seasons are made palpable through the presencing of different natural things—"roots, birds, eggs, edible grubs, lizards" (Green 1984, 10-11), registered multi-sensorially by individuals through their powers of sight, touch, taste, smell, and sound.

The Benedictine monk, Dom Rosendo Salvado (1814–1900), who established the New Norcia monastery on the banks of the Moore River north of Perth, commented that "it seems that some natives divide the year into six different seasons; but many others divide it into four, which they call *cielba [jilba], mocur, ponar, piroc*, that is, autumn, winter, spring, and summer. The months are distinguished from one another by the moon, but they are not given individual names, or divided into weeks. Again the days are not distinguished except by the position of the moon" (Salvado 1977, 131). Curiously, Salvado only references four of six Albany seasons, *Jilba (cielba), Mokkar (mocur), Pourner (ponar)*, and *Birok (piroc)*, despite the existing account of colonial doctor Isaac Scott Nind (1797–1868), published in 1831. Nind (1831/1979, 35) notes that "the greatest assemblages [of Albany area Nyoongar people] are in the autumn (*pourner*), when fish are to be procured in the greatest abundance." He observed six seasons "beginning with June and July, or Winter: *Mawkur, Meerningal, Maungernan, Beruc, Meertilluc,* and *Pourer*" [italics added] (Nind 1831/1979, 54). Salvado's emphasis on the four Nyoongar seasons might reflect an intractable four-season logos that simply could not rationalize liminal states of temporality for, as he says, "it *seems* that some natives divide the year into six different seasons [italics added]." Moreover, Salvado noted that Nyoongars reckoned weeks and days according to the moon, but that these smaller denotations of time were not as important as the six seasons in the Nyoongar temporal order.

In contrast to the Nyoongar bush calendar, as the previous section demonstrated, the Gregorian calendar pivots on the precise calculation of time in determining the four seasons and the exact placement of Christian holy days. In his discussion of Aboriginal temporality, Mike Donaldson (1996a, 193) describes a non-Western sense of time as "enveloping. Both cyclical and circular, it accorded with the need for seasonal movement, the aggregation and disaggregation of groups." "Nyoongar time," for Donaldson (1996a, 200-1), reflects "close ties with the land [...] which blurred the distinctions between work and leisure." Based on this temporal sense, Nyoongar seasons reflect natural events and are connected to the procurement of food and movements of communities—thereby bridging ecological and structural time. Salvado (1977, 289) observed "it is worth noting that the Australian natives [...] use the title 'grass season' of the period in which the new grass is born and the buds open, that is, the months corresponding to April–May of the northern hemisphere (our months, however, being autumn for them)." In the Perth-area Nyoongar calendar, the months of April and May correspond to *Geran*, signified by the coming forth of grass buds and associated flora and fauna. The budding of grass is an important ecological phenomenon in the annual cycle of the *kwongan* sand plain ecosystem fringing Salvado's New Norcia settlement to the west. Salvado's statements suggest that *Geran* is a shifting denominator—a movable category of time—that depends on a variety of biotic, abiotic, astronomical, and cultural factors, rather than the pre-set calendrical months of April or May, occurring in a predetermined fashion on the first and last days of their cycles.

As embodied temporality (my preferred term used here to encompass ecological and structural time), the seasons governed traditional Nyoongar movements, activities, and customs. An article in an 1833 edition of *The Perth Gazette* noted that a reconciliatory meeting between warring settlers and Nyoongars "could not be effected at present, as the tribes were so much dispursed [*sic*], and not until the *yellow* season (the bloom of the Banksia,) in December, January, and February. At this time the country is generally fired [italics in original]" (The Perth Gazette 7 September 1833, 142). The three months listed in the article correspond to *Birok* and *Burnoru* when different species of banksia bloom, including the bull banksia (*mangite* or *Banksia grandis*)—the flowers producing an abundance of nectar, which was steeped in water or sucked directly by Aboriginal peoples. During the yellow season, the Christmas tree (*Nuytsia floribunda*), known as *mudja*—the Nyoongar word for fire (Ryan 2012a, 79-

81)—also blossomed. Further to the color symbology of the seasons, it was during *Birok* and *Burnoru* that Nyoongar people set fires to encourage grazing animals and the regrowth of food plants (Hallam 1975). However, banksia nectar—as a sensuous thing announcing the seasons phenomenologically—was also important during other times of year. Writing in October 1833, George Fletcher Moore (2006, 292) reported that "this is the season now for young parrots. I am told that the natives suck the honey out of their bills which the mother has just fed them with from the Banksia flowers." During the *Kambarang* season, Nyoongars hunted young birds and eggs (Elkin 1943, 36). Additionally, Moore (2006, 315) from March 1834 observed that Nyoongars "pull the blossoms of the red gum tree (now in flower), steep them in water, and drink the water, which acquires a taste like sugar and water by this process." Between *Burnoru* and *Makaru*, the red gum tree (*marri* or *Corymbia calophylla*) flowers throughout the South-West region.

As suggested by the term "the yellow season," some contemporary explanations of the Nyoongar seasons point to color typologies with phenomenological bearing on human perception of the temporal world. However, these typologies are not always consistent between sources. *Our Place Newsletter* (Kurongkurl Katitjin Centre 2011, 5) notes that colors are used to teach seasonal knowledge and to help Nyoongar people identify "the correct time of year" for certain activities. *Birok* is associated with the color red or *mirda*, symbolizing heat, fire, and the sun. *Burnoru* is signified by the color orange or *yoornt mirda*, representing the profusion of fish and lack of rain characteristic of this season. *Geran's* color is green, or *nodjam*, correlating to the return of cooler weather and the light green appearance of eucalypt trees. *Maggoro* is blue, or *wooyan*, with dark blue, specifically signifying the onset of rain and cold temperatures. *Jilba* is associated with the color pink, or *mirda mokiny*, with pink or purple indicating the proliferation of wildflowers in the South-West during this season. Finally, *Kambarang* (not *Birok* and *Burnoru*, as indicated above) is linked to the color yellow, or *yoornt*, symbolizing the arrival of hot weather and other "yellow" events that complete the yearly cycle (Kurongkurl Katitjin Centre 2011, 5).

As a contemporary teacher of Nyoongar seasonal knowledge, Len Collard, a Traditional Owner of the Whadjuck or Perth metropolitan area Nyoongar, comments that "we utilize six seasons of the year for food and sustenance, and never damage or kill our resources unnecessarily. The land is our mother and our nurturer and our guiding light" (Stasiuk and Sillifant 2005). Collard links the six Nyoongar seasons to meteorological conditions (ecological time), such as wind directions and temperatures, but also to the procurement of food, maintenance of cultural knowledge, and performance of ceremonies (structural time):

> The Nyoongar seasons are Bunuru with hot easterly and north winds [...] *Djeran* becomes cooler with wind from the south-west [...] *Makuru*, cold and wet with westerly gales [...] *Djilba*, becoming warmer [...] *Kambarang*, rain decreasing [...] *Birak*, hot and dry with easterly winds during the day and south-west sea breezes in the afternoon [...] There were between 30 and 40 distinct roots, nuts, and vegetables eaten by Nyoongar, which are gathered nearly all-year round [...] There was hardly any shortage of food throughout the six season cycle with *katitjin*

or knowledge given to the Nyoongar by the *Waagal* [Creation Serpent] to manage our land according to the seasons. (Collard in Stasiuk and Sillifant 2005)

Hence, for Collard, the endemic six seasons of the Nyoongar derive uniquely from the meteorological, botanical, and cultural contexts of the South-West Australian landmass. Such variables factoring into the Nyoongar bush calendar coalesce to designate the onset of each of the seasons. Crucially, however, the physical openness of humans to the sensuous nuances of experience is the mode through which the things of the South-West—red gums, banksias, the wind, roots, nuts, vegetables, nectar, birds—announce themselves. This phenomenological mode of gaining seasonal knowledge shifts, not only from season to season, but from region to region. Thus, a phenomenology of the seasons engages people and the things of nature in their milieux of dwelling, leading to place-based and embodied temporality.

Seasons of Our Dwelling: The Indigenous Weather Knowledge Project (IWKP)

In December 2012, Edith Cowan University and the Bureau of Meteorology launched the Nyoongar weather calendar as part of the Indigenous Weather Knowledge Project's continuing effort to preserve and promote traditional Australian Aboriginal seasonal knowledge (Edith Cowan University 2012). The online calendar lists the six Nyoongar seasons as *Birak, Bunuru, Djeran, Makuru, Djilba,* and *Kambarang* (Australian Bureau of Meteorology 2010). (See http://www.bom.gov.au/iwk/). As an educational and heritage-based tool, the IWKP website emphasizes that human perception is fundamental to understanding the endemic South-West seasons: "the Nyoongar seasons can be long or short and are indicated by what is happening and changing around us rather than by dates on a calendar" (Australian Bureau of Meteorology 2010). For example, *Birak* (December–January), the first summer or the season of the young, is marked by the easing of rain, the onset of warm weather, sea breezes from the southwest, easterly winds, fledgling birds, reptiles shedding their skin, and baby frogs. *Bunuru* (February–March), the second summer or season of adolescence, is signified by high heat and little rain, hot easterly winds, the white flowers of jarrah, marri, and ghost gums, and the bright red cones of female zamia (*Macrozamia riedlei*). *Djeran* (April–May), the ant season or season of adulthood, features the breaking of hot weather, cooler nights, light breezes from the south-east or south-west, flying ants, the red flowers of *Corymbia ficifolia* and *Beaufortia aestiva*, and the flowering of other banksias. Traditionally, during *Djeran*, Nyoongars consumed zamia nuts that were, earlier in the year, stored underground or water to hasten the food crop's detoxification. Shelters known as *mia-mias* were repaired in preparation for the coming cold season.

In addition to the Nyoongar calendar, the IWKP outlines the endemic seasonal knowledge of eight other Aboriginal cultures: Brambuk, D'harawal, Walabunnba, Yanyuwa, Jawoyn, Miriwoong, Wardaman, and Yawaru. For example, in the Yawaru calendar situated north of Broome, Western Australia, uses ecological indicators, such as the ripening of the cocky apple and the availability of wild yams, to indicate *Mankala* or the wet season (Australian Bureau of Meteorology 2010). The Walabunnba people, living approximately 300km north of Alice Springs in central Australia, recognize two seasons: *Wantangka* (the hot weather) and *Yurluurrp* (the cold weather). During *Wantangka*, the sweet bush plum is eaten when the fruit turns dark, and special "hot weather" ceremonies are performed during the season.

The IWKP website is the outcome of a collaboration between Indigenous Australian communities, the Aboriginal and Torres Strait Islander Commission (ATSIC), the Bureau of Meteorology, and Monash University's Centre for Australian Indigenous Studies (CAIS) and School of Geography and Environmental Science. The South West Aboriginal Land and Sea Council granted permission to the IWKP to display culturally sensitive Nyoongar seasonal information. Moreover, the IKWP is integral to the Bureau of Meteorology's *Reconciliation Action Plan 2012-2015*. One of the plan's objectives is "to liaise with community elders to expand traditional knowledge of weather and climate through seasonal calendar information" (Australian Bureau of Meteorology 2012b, 2). Indeed, as one of the original institutional advocates of the Western seasons in Australia, the Bureau of Meteorology concedes that "the four seasons we've adopted are not entirely appropriate for all regions of Australia." The Bureau praises "natural calendars" or "bush calendars" for reckoning the "natural seasons" according to ecological phenomena, such as "fruits, blossoms, insects, animals, as well as the temperature and whether it was a wet time of year" (Australian Bureau of Meteorology 2008b, 10).

In foregrounding ecocultural (environmental and cultural) way marks which herald the passage of time, such as the ripening of the bush plum, the IWKP offers a phenological tool for appreciating the Aboriginal Australian seasons. The word *phenology* stems from the Latin *phaeno* and the Greek *phaino*. *Phenology* and *phenomenon* share a common etymological root in the Greek *phainein* for "to show" (Harper 2012b), from which the words *phantasm* and *phenotype* come. Introduced in an 1853 article by Belgian botanist Charles Morren (1807–1858) and advanced in the 1880s by the Austrian botanist Karl Fritsch (1864–1934), phenology can be defined as "the study of periodic biological events in the animal and plant world as influenced by the environment, especially temperature changes driven by the environment" (Schwartz 2003, 3). For Morren, *phenology* meant "to show, to appear: the science of phenomena that appear successively on the globe" (Keatley and Hudson 2010, 1). The first published English definition of phenology, following the term's adoption by the Council of the Meteorological Society in 1875, read "the observation of the first flowering and fruiting of plants, the foliation and defoliation of trees, the arrival, nesting, and departure of birds, and such like" (Anon. 1884 quoted in Keatley and Hudson 2010, 1). Further along, a 1972 American committee on phenology employed the following definition: "the study of the timing of recurring biological events" (Leith quoted in Keatley and Hudson 2010, 2).

Forwarded by the Bureau of Meteorology, the IWKP intersects with scientific knowledge of weather and the seasons. The project also highlights the seasonal heterogeneity of Australia. As an online resource containing an evolving collection of phenological information about Australian endemic seasons, the IWKP provides catalogue-like indications of the first occurrences of ecocultural events in the respective regions of the nine Aboriginal societies featured. However, immediate experience through the senses—of the endemic things of Australian places, including bush plums, zamia nuts, and wild yams—is indispensable to comprehending the seasons in their sensuous presencing. The phenomena that indicate seasonal passage—archived by the IWKP in consultation with elders—are integral to the habitus of people in a place. In a scientific sense, a phenology functions, in part, like a compendium of events. In contrast, a phenomenology of the Australian seasons thus integrates biotic, abiotic, cultural, cosmological, ceremonial, ontological, and corporeal aspects—all essential to Aboriginal temporal orders. Fostered through a phenomenological foundation, immediate embodied knowledge of the seasons is a much-needed complement to the largely cognitive, visual, and events-based perception of temporality. For example, although the bush plum flowers in summer, if I have never seen, tasted, or smelled its fruit, my knowledge of the endemic seasons of where I live will be limited—I must engage with the things of the seasons through which events manifest. I must be with the seasons. Hence, endemic seasonality—whether *Mankaru* of the Yawaru calendar or *Djeran* of the Nyoongar calendar—is intimate with embodied temporality, the recognition of time's passing through corporeal involvement with one's place. In sum, the IWKP offers a groundwork for phenomenological exploration of the seasons. Nevertheless, the information preserved and promoted by the IWKP should not be regarded as a substitute for immediate, real-time encounter with seasonal things—but rather as a fact-based catalyst to such encounters.

Although an incomplete digital tool in itself, the IWKP offers an accessible template for engaging with the Australian seasons through sensory experience. In analyzing the IWKP in these terms, I have distinguished between a phenology of the seasons—as cognitive recognition of temporal events—and a phenomenology of the seasons—as embodied, immanent interaction with the seasonal things that herald such events. Ultimately, a phenomenology complements a phenology of the seasons, here encapsulated by the IWKP. Returning to Entwistle's proposal, even the more considered kinds of seasonal paradigms risk imposing a managerialist grid on the plural landscapes—bioregions, places, locales—that comprise Australia as a highly diverse ecocultural whole. For instance, in the five-season scheme, spring as a temporal denomination is entwined with the flowering of native plants. Although botanically sensitive, this prioritization backgrounds the other physiological events in the annual cycles of flora—as well as the cultural, sensorial, spiritual, ethnozoological, astronomical, and climatic considerations that collectively signify the seasons (Clarke 2009). Ecological and structural time lived out in place synergetically infuse the experience of embodied temporality. Flowering phases reflect one aspect of an endemic (land-based or Indigenous) calendar as an environmentally and culturally integrated whole. Through the fusion of phenomenological (experiential, sensory, place-based, actual) and phenological (cognitive, visual, enumerative, digital) approaches, the endemic seasons of Australia can be appreciated in their depth and extent.

Conclusion: Living with Seasonal Plurality in Australia

As suggested by Entwistle's call, Australia has an uneasy relationship to the four seasons of the Western calendar and the northern hemisphere. The rethinking of the Australian seasons entails the recognition of a multiplicity of seasons, calendars, cultures, and places. National models—whether the Gregorian four seasons or Entwistle's proposal for five seasons—can co-exist dynamically with robust traditions of endemic seasonality, exemplified by the Nyoongar six seasons and the Indigenous Weather Knowledge Project. As Clarke (2009, 101) comments, "while increasing globalization prevents European Australians from rejecting the European-derived calendar in favour of a plethora of regional calendars, the future investigation of indigenous seasonal knowledge and behaviour offers to help develop more relevant approaches to landscape management." As I have argued in this chapter, the "future investigation" of Indigenous calendars will need to be experiential, sensory, and place-based.

Hence, a dialogical perspective on the seasons is phenomenological and phenological—cognitive and bodily—comprising the proposed portmanteau *phen(omen)ology*. Attending to the seasonal things of place which pronounce the passage of time, a phen(omen)ology is a reflexive perspective on the seasons that blurs the distinction between intellection and embodiment. Moreover, a phen(omen)ology recognizes that actual seasonal boundaries vary year to year and from place to place according to an array of ecocultural factors. As Heidegger acknowledged, "a boundary is not that at which something stops but, as the Greeks recognized, the boundary is that from which something *begins its presencing* [italics in original]" (1971, 152). Learning to be with seasonal things requires knowledge of when ecological events tend to happen coupled to immediate sensory witnessing of their manifestations.

Chapter 2: Stories of Snow and Fire: The Value of a Pluralistic Environmental Aesthetic

Introduction

Narratives enable us to appreciate the natural world in aesthetic terms. Firstly, narratives can galvanize for the reader or listener a sense for another person's experience of nature through the aesthetic representation of that experience in language. Secondly, narratives can encode and document for the human appreciator as writer an experience of nature in aesthetic terms. Through different narrative lenses, the compelling qualities of environments can be crystallized for both the reader (who vicariously experiences nature through language) and the human appreciator (who directly experiences nature through the senses). However, according to philosopher Allen Carlson's "natural environmental model" of landscape aesthetics, science provides the definitive narrative that represents nature on its own terms and catalyzes appropriate appreciation. In this chapter, I examine Carlson's claim and argue for an environmental aesthetic philosophy of narrative multiplicity. Such a model would draw from scientific, Indigenous, and journalistic narrative modes toward a critically pluralistic environmental aesthetic of the natural world. The ethical framework I propose— the function of which I characterize simply as narrative "cross-checking"— acknowledges the value of narrative heterogeneity in expressing and generating aesthetic experience of environments. The position I put forward in this chapter is developed through extensive treatment of these three narratives expressed within two examples, one of geographical place and one of environmental practice. As I will suggest, Denali, the prominent Alaskan mountain, can be aesthetically appreciated through the narrative forms enumerated above. As a second case study, the traditional burning regimes of Indigenous peoples reveal collectively how a critically pluralistic environmental aesthetic of narratives can be applied to—and identified to exist within—ecocultural practices, such as the seasonal firing of the landscape.

Stories of Snow and Fire

Denali presides above the Alaskan interior, its ermine coat of snow contrasting with the verdant basin below (Figure 2). The angularity of its bulk contrasts noticeably with the undulating green valley of the foreground. Unlike higher counterparts in the ranges of Nepal that compete for prominence in the skyline, Mount McKinley (the European name for Denali) protrudes starkly from the Alaskan flatlands as a kind of solitary monarch of the subarctic Alaskan vista. On a radiant early summer day, Denali's glaciers glisten—opaque, milky ice interspersed with dark skeins of granite. To consider Denali as merely a scene framed for our sensual pleasure, however, is to dismiss the manifold aesthetic experiences that the natural world—including mountainscapes—affords us. Indeed, aesthetic experience of the environment is generated by heterogeneous human faculties, including the bodily senses of touch, taste, and smell; our emotion, imagination, and intuition; as well as scientific, Indigenous, and folkloric knowledge of places, such as the mountain Denali.

Indeed, the immediate visual characteristics of the mountain—the glistening snow and the angularity of rock—can serve as a picturesque basis for a more nuanced, layered, and sensorial mode of appreciation. For instance, through our imagination, we could project into time and attach possibilities to the mountain; envisioning summer brings greater attention to the rock features underlying the snow pack, along with the wilderness background and pellucid blue sky. Moreover, we could physically encounter the mountain, calling on other sensory faculties and stimuli—for example, the numbing chafe of wind on our cheeks; the smell of evergreens; or kinetic sensations of muscle fatigue and accelerated heart rate. Additionally, in heightening our aesthetic experience of Denali, we could reflect on its natural history: the mountain's narrative detailing the geological events responsible for the aesthetic qualities of the scene it imprints within our perception. For the purposes of this chapter, it is the role of scientific, Indigenous, folkloric, and popular narratives as catalysts for aesthetic experience of the natural world that forms the core of a critically pluralistic environmental aesthetic based on narrative heterogeneity.

Comparably, in the boreal forests of northern Alberta, anthropologist Arthur Bailey, as depicted in the film *The Fires of Spring* (1978), stands beside train tracks that strike through the landscape (Lewis 1978a). Prairie parkland, punctuated with willow and aspen, unfolds to the west. This is the land beheld in 1903 by the drafters of Canadian title to this region, a boreal mosaic of grasslands—forest studded with patches of muskeg—whose productive soils attracted ambitious European settlers. This is also the land frequently burned by the railroad authority to maintain unobstructed train lines—a scenario carried out in many places in the world (for example, Lunt and Morgan 2002). To the western side of the tracks, a different situation prevails: aspen forest dominates the plant community where fires have been absent for several decades. In fact, the boreal forest of northern Alberta had an open configuration before provincial bans on burning were enacted in the early 1900s (Lewis 1978b). Traditional firing in the spring induced an early spurt of succulent herbs and shoots, and maintained a more uniform growth of the prairie, creating the preferred habitat for the animals on which the Indigenous Dene people subsisted. Monikers like "High Prairie," however, are now anachronisms, hinting of a time when frequent, light spring burning maintained an expansive prairie mosaic. The muskeg and forest pattern, which also provided grazing for moose and other herbivorous animals, has been altered by unconstrained growth of brush and trees. Importantly for the focus of this chapter, the aesthetic qualities of the prairie landscape have been altered by the exclusion of burning practices.

Figure 2. Denali is the highest peak in the United States. Its name comes from the Koyukon Athabaskan language for "The High One." The mountain has an elevation of 20,320 feet (6,194 m) above sea level.
Source: National Park Service, Public Domain, Wikimedia Commons.

The traditional burning of an ecosystem is a significant component of ecological management in Indigenous practices across the world today. Historically, burning has been a vital survival strategy with a broad range of purposes, the most salient of which is to ensure continued access to subsistence resources. Frequent, low-intensity, and patchy firing of the land, in conjunction with restrictions of such practices in select areas, influenced the composition of ecosystems by promoting the growth of resource plants and inhibiting the predominance of less desirable plant species. Examples from North America, Australia, and Africa collectively express the technological sophistication of Indigenous management of the land through a burning regime implemented and designed specifically for the weather, season, biota, and culture of a place. The termination of traditional fire management systems has resulted in discernible alterations in biotic composition through the suppression of fire-sensitive species and the encouragement of fire-promoting plants, such as grasses. The relatively new field of fire ecology provides us a wealth of knowledge and perspectives for understanding the distinctive cultural ecology of human-made fires (Lewis 1978b, 402). The incorporation of traditional fire technology into contemporary ecosystem management paradigms has become increasingly commonplace.

Yet, just as our appreciation of Denali is enhanced by the many stories surrounding the mountain, the emerging field of fire ecology could benefit from ongoing engagement with the diverse narratives of Indigenous burning practices, not only those forwarded by managerialist conservation science. Indeed, some of the narratives of Indigenous burning practices involve spiritual and cultural meanings linking landscapes and people through concepts of stewardship. For example, in Australia, as anthropologist Deborah Bird Rose observes in *Nourishing Terrains*, "when Aboriginal people speak English, they describe their burning practices as 'cleaning up the country'. There is a well-defined aesthetic; country which has been burned is country which looks cared for and clean. It is good country" (Rose 1996, 65). Hence, the story of the burned landscape, for Aboriginal Australians, is closely linked to an ecocultural condition involving long-term regeneration over seasonal cycles. In a traditional sense, caring for country is connected to an appropriate aesthetics of burning, ensuring the long-term viability of the land. Moreover, Aboriginal stories of burning the landscape also implicate the spirits of deceased ancestors; and hence narratives of fire are also narratives of family and kin. An aesthetics of caring for country, therefore, is a broader ecocultural aesthetics based in stories. John Bradley reports:

> For the Yanyuwa the burning of country is an important way of demonstrating a continuity with the people who have died, their ancestors, or li-wankala, 'the old people'. The spirits of these people are said still to inhabit the landscape; they still hunt, sing, dance and are said even to still burn the country. Indeed it is spoken by the contemporary old people that before the coming of the white people, the spirits of the deceased kin would set fire to the country themselves for hunting, and up to quite recently, country that was burnt was left for several days so the spirits of the deceased could hunt first. (quoted in Rose 1996, 71)

Bradley goes on to say that "country that has not been burnt for a long time is described as being 'shut up'. Visually this can be seen by the increase in the understory vegetation, and on the islands by the increase of choking vine thickets" (quoted in Rose 1996, 72). The park-like, picturesque appearance of the Australian landscape in many places, as noted in the journals of numerous nineteenth-century visitors and explorers, resulted from the burning regimes of Aboriginal societies. Considering the meeting of both narratives, we see the convergence of a techno-scientific European aesthetics of colonial landscapes and a regenerative Indigenous aesthetics of caring for country as "nourishing terrain."

Stories, such as these provided by scientists, anthropologists, and Indigenous peoples, satisfy our collective needs for understanding what we perceive and experience in nature. As such, stories can be enduring foundations for the appreciation of the natural world. Moreover, narratives (used in this chapter interchangeably with the word *stories*) can underlie the qualities and content of our eco-aesthetic experiences. Narratives balance and moderate immediate, highly subjective, sensory reactions to aesthetic features, such as the jaggedness of Denali or the burning of the Australian landscape, ensuring that our appreciative responses to the natural world are appropriate and translatable. Three narrative forms—scientific, Indigenous, and popular/journalistic—will be discussed through the examples of snow and fire. I argue that the mechanism of "cross-checking" between narrative forms is indispensable for establishing the ethical appropriateness of our aesthetic responses. Narrative heterogeneity, as I will go on to outline, is the core of a critically pluralistic environmental aesthetic. A broad basis for an environmental aesthetics emerges when different narrative modes commingle. The purpose of this article is to point to the need for narrative commingling. Guiding this discussion, Denali and the firing practices of global Indigenous cultures will provide tangible examples of how narrative heterogeneity can deepen appreciation of the natural world as part of a sensorially refined, ecologically appropriate, and broadly attuned human aesthetic. I acknowledge that my example related to snow focuses on the limited area of Denali while my examples of fire address Indigenous burning practices from around the world. Indeed, a more balanced presentation of evidence would integrate narrative examples from other prominent and well-storied peaks around the world, such as Mount Everest and Mount Kilimanjaro. Despite this limitation, which should be addressed in further studies of philosophical aesthetics, in broad terms an integrated approach to environmental narratives is significant because it responds to the current context of climate change in which conflicting accounts populate the media—from scientific corroboration of broad-scale climatic perturbations to the emotionally-charged voices of climate change skeptics; and from the place-based experiences of Indigenous peoples in polar regions facing environmental threats to their small settlements to the incomprehensible scale of atmospheric transformation and the demise of planetary life-support systems. More specifically, increasingly frequent fires in the arid regions of the United States and Australia, along with the shrinking of Alpine glaciers in Europe, heighten the implications of these two case studies of ecology and the senses: of snow and fire.

Science as Narrative: Carlson's Natural Environmental Model

Science makes available one of many narratives for aesthetically contemplating Denali and seasonally burnt landscapes in North America and Australia. Specifically, the mountain's geological story articulates those forces and events that coalesce in its present topographical form. Geology explains that Denali is primarily granitic and consists of igneous rock—forming a batholith—that collided with various plates during the formation of North America. The uplift of this batholith resulted from continued tectonic collision. The high altitude of Denali results from its position at the boundary between the Pacific and North American plates. Here, subduction, where one plate goes under the other, and strike-slip motion both occur and continue to propel Denali upward. Of course, much of the scenery in Denali, rather than static, is presently being shaped by a number of active glaciers and other dynamic geological forces (Schwartz n.d.). In sum, geology presents the aesthetic (human) subject with a narrative expressing the dynamism of the natural world over time. The object of aesthetic concern— the mountain—is linked to companion objects in the landscape (other mountains, water bodies, flora and fauna) through the narratives of natural history.

Through environmental stories, we come to know the characters and protagonists (the massive batholiths, the even more colossal plates, and the glaciers, as well as our central protagonist, Denali) and the dramatic forces shaping their interactions (the titanic force of plates colliding and the strident scraping of ice against granite). Consequently we can experience the *denouement* of a kind of plot tension: the relatively abrupt uplifting of Denali into its present high-altitude position. An aesthetic appreciation of Denali stems from its characterizations within the geological narrative: the massiveness of the plates; the somber, heavy tone of the word "batholith" itself; and the deep sense of geological time involved in all of these complex transactions. In order to forge a pluralistic environmental aesthetic, elements of the geological narrative would intermingle with my subjective perception of Denali's aesthetics to produce a holistic response—one which is representative of Denali as a living place. In such a model, science provides the story which enhances—rather than dictates, predetermines or dominates—our regard *for* and appreciation *of* the mountain.

Allen Carlson purports the invocation of science as the single "appropriate" narrative structure for aesthetic appreciation of the natural world through his "natural environmental model." As the essential narrative for an environmental aesthetic in Carlson's view, science "is a paradigm of that which reveals objects for what they are and with the properties they have" (Carlson 1993, 219). Carlson asserts that Mannison's position—nature does not have a capacity for aesthetic appreciation because it lacks the artistic design or intention of an artist—is only true where the design approach of paradigmatic art appreciation is imposed on environmental aesthetics. According to Carlson's argument, awareness of Denali's natural history narrative logic is an underlying constituent of our appreciative response.

Here, Carlson discerns between "design" and "order" appreciation. He argues that the aesthetic experience of the natural world must break free of design theory in which an object is regarded for its immediate phenomenological presence, apart from the history it bears. In his view, such an approach tends to exclude relationships between animate and non-animate objects in landscapes. Carlson moves toward "order appreciation" by describing avant-garde and anti-art works such as Jackson Pollack's drip paintings (spontaneity and chance), Duchamp's Dadaist "found art" (emphasis on the selective process of the artist) and Dali's Surrealist painting (juxtaposition of disparate images, colors, and textures). These works, for Carlson, prioritize the story behind the object over the design principles exhibited through the object; for instance, the artistic essence of a toilet bowl on display reflects the reasoning of the artist encapsulated in a narrative, comprising where the object was found, why it was chosen and perhaps what political or intellectual events the artist responded to. Carlson terms this "order appreciation." In contrast to design appreciation, these works exhibit an ordered pattern and thus their appreciation will be different: "The object embodying the design no longer embodies design [but the discerning of order] and the individual who embodies the design is no longer a designer [a selector or distinguisher, instead]" (Carlson 1993, 221). According to Carlson, avant-garde artworks, by embodying order appreciation, present a possible model for an environmental aesthetics through scientific narrative.

Carlson proposes that, in the appreciation of order, we should look for the narrative behind the genesis of the work. While the story behind an ordered artwork, such as Pollack's drip painting, is the particular technique or circumstances inscribed within it, an analogous framework for environmental aesthetics is natural science. The natural order of science makes the pattern comprehensible and appreciable, as design does for a designed object. The objects or landscapes embodying the order are informed by environmental forces. The relevant vitalities of nature are analogous to the artist's narrative setting out how or why an artwork came into existence. For Carlson, science is the essential narrative for aesthetic experience of nature; through extrapolation, the authenticity of one's appreciation of ordered environmental phenomena depends upon one's understanding of and engagement with natural science.

Hence, the "natural environmental model" prioritizes scientific knowledge in the aesthetic judgment of the natural world. As such, the natural historian assumes the position of the artist in creating ordered artworks by revealing the story behind the creation. Moreover, science is emphasized over other narrative forms because of its inherent objectivity. Furthering Carlson's argument, Marcia Eaton supports the exclusivity of science as the appropriate narrative for natural aesthetics: "In the case of aesthetic appreciation of nature, the scrutiny is based upon and enriched by scientific understanding of the workings of nature; without that one cannot be certain that one's response is to nature and not to something else" (Eaton 1998, 149). The connection between science, objective knowledge of nature, and truth formation prompts cognitive theorists such as Carlson and Eaton to extend *a priori* epistemological footing to disciplines like geology and fire science in countering subjective falsehoods—those assessments deemed irrelevant or inappropriate to aesthetic appreciation of the natural world.

Returning to Denali: Indigenous Cosmology as an Appropriate Narrative for Aesthetic Appreciation of Nature

The Koyukon Athapaskan, who can trace their presence in the northwest interior of Alaska 1,500 years back, explain the presence of Denali in their corpus of stories called the "Distant Time:"

> The Raven, incarnated as a young man, had paddled his canoe across a great body of water to ask a woman to marry him. She refused to be his wife, so he made her sink into the mud and disappear; and then he began paddling back home. The woman's mother kept two brown bears, and in her anger she told them to drown the young man. They dug furiously at the lake's edge, making huge waves everywhere on the water. But Raven calmed a narrow path before him and paddled on. Eventually he became exhausted, so he threw a harpoon that struck the crest of a wave. At that moment he fainted from the intensity of his concentration, and when he awoke a forested land had become a small mountain. Then it had glanced off, eventually striking a huge wave that solidified into another mountain—the one now called *Deenaalee*, or Mount McKinley. (Nelson 1983, 34)

This Koyukon narrative of Denali's genesis points to the importance of Indigenous knowledge in aesthetic appreciation. A listener could become enthralled by the image of a wave solidifying to become a mountain. The image and story account—as successfully and appropriately as natural science—for the cascading features of the mountain, waves frozen in a tense energetic moment before crashing to the lake. Raven's spear becomes a protracted line of rock extending from the summit. For a critically pluralistic aesthetic, these images of ice and cascades are sufficient enough to initiate an aesthetic response to the mountain. Thus, the success of this kind of Indigenous narrative in stimulating an aesthetic experience contests Carlson's claim that natural science, and its commonsense predecessors and analogues, have unequivocal standing over aesthetic appreciation of the natural world.

Thomas Heyd demonstrates that there are key problems with the "natural environmental model." He argues that aesthetic appreciation should benefit from diverse narratives, developed by comparably diverse cultures across the globe. Moreover, for Heyd, scientific knowledge can be counterproductive to our aesthetic experience because "it directs our attention to the theoretical level and the general case, diverting us from the personal level and the particular case that we actually need to engage" (Heyd 2001, 126). Heyd explains that scientific knowledge, in its capacity to universalize, can fail to account for particularities of a specific natural object or landscape, such as a mountain or a landscape managed with fire. However, these particularities are essential for an aesthetic response most representative of a certain tree, mountain, or ecosystem. For instance, in the case of Denali, the geological narrative fails to express the subtle, human-scale chasms and protuberances—those aesthetic characteristics which distinguish Denali from other mountains on the planet.

Proponents of the exclusivity of scientific narratives tend to discredit the role of Indigenous knowledge in the aesthetic experience of the environment. The alleged incompatibility between scientific understanding and Indigenous knowledge reflects a modern dichotomy over the existence of truth, which, crudely put, permeates many facets of Western (and Westernized) cultures and extends into environmental aesthetics. Anthropologist Eugene Hunn comments that "in America today a 'myth' is a species of falsehood. For Indians [Native Americans], myths were and are a species of truth" (Hunn 1990, 85). The disregard for Indigenous cosmology and other forms of "folkloric" knowledge as a legitimate basis for aesthetic experience, is part of a broader marginalization of non-scientific knowledge, and, conversely, a fuller investment in scientifically-dictated epistemology. Environmental philosopher Holmes Rolston, although he doesn't overtly discredit folkloric knowledge, considers the custodial role of science over myth: "Science is necessary to banish ('deconstruct') these myths, before we can understand in a corrected aesthetic" (Rolston 1995, 382). Similarly, Carlson dismisses folkloric and Indigenous knowledge in favor of science, although he does so tactfully and tacitly: "Thus, perhaps nature is easiest to appreciate when our account of it is simplistic anthropomorphic folklore: a story of almost human gods or godlike human heroes not unlike the so-called primitive peoples who are said to feel exceptionally close to nature [...] to appreciate nature is to confront either an almighty god or blind natural forces by way of ourselves [...] it is no surprise if typically only the confrontations with nature are marked by overwhelming wonder and awe" (Carlson 1993, 223). For Carlson, mythical creatures are not credible; folklore is an indirect, culturally-specific and therefore less verifiable (universalized) source for experience of the natural world. Again, science is the primary basis for aesthetic experience. For these theorists, science makes possible a genuine natural aesthetic through objectivity and freedom from non-scientific cultural imagination.

Heyd acknowledges the criticism of folklore as subjective, culturally-based, and value-driven in contrast to the supposedly objective, universal, and value-free tenets of science. He argues that the credibility of cosmological entities is irrelevant to the capacity of such stories to guide and modulate the aesthetic appreciation of the natural world. For Heyd, stories like the Dreaming of Australian Aboriginal peoples—because of their specificity and concreteness—pull us, so to speak, into the natural scene, object, or event, in a manner equally compelling to science: "[the Aboriginal Dreamtime stories illustrate] well the details in the landscape that may become perceptually salient through knowledge of it, much in analogy to the manner in which a rock face might become perceptually salient for someone knowledgeable of the geological story concerning its different strata" (Heyd 2001, 133). As Heyd claims, scientific abstraction and universalization can divert our serious consideration of the inimitable truth of certain natural phenomena. Yet, the narratives of Indigenous cultures can channel our attention to the uniqueness of that natural phenomenon, endemic to the geographical region of the ecocultural milieu that generated the story. Understanding the limitations of all narratives, Heyd offers the following caveat: stories need to be evaluated on an individual basis for the degree to which they accentuate or obfuscate aesthetically appreciable characteristics of nature.

Yuriko Saito extends moral terms to Heyd's principle of narrative multiplicity. For Saito, morally appropriate narratives are those that acknowledge the reality of nature apart from human society, thereby producing a fair representation of nature. In Saito's view, the power of folklore to engage aesthetic appreciation of the natural world depends on the "degree" of the narrative. Creation myths, as universalistic narratives, attempt to account for the existence of the whole Earth, and do not provide a suitable framework for the valuation of specific mountains, rocks, or animals. For Saito, these are morally unacceptable because they fail to reflect the individual history of natural features appropriately. Bioregional narratives, on the other hand, represent the polar opposite side of the narrative spectrum where specific attention can be given to individual attributes of objects. Such stories, exemplified by the Koyukon myth of *Deenaalee*, have the moral capacity to provide an appropriate environmental aesthetic according to the degree to which they express the bioregional history of the aesthetic "object." Saito explains that "the more specific the observation and attention become (as in bioregional narratives), the more sensitive we are to the diverse ways in which natural objects speak about their respective histories and functions through their sensuous qualities" (Saito 1998, 148). Across diverse narratives, for Saito, the framework guiding what constitutes a morally appropriate environmental aesthetic is the degree to which the story represents nature on its own terms. Saito's terms constitute a bioregional framework for narrative environmental ethics.

As with Ice, So with Fire: Science and Stories of Burning in Aboriginal Australia

The discussion of Denali and the appreciation of its aesthetic qualities points to a critically pluralistic environmental aesthetic encompassing science and Indigenous cosmologies alike. I now turn to a treatment of the diverse narratives that can inform the appreciation of fire, beginning with scientific interpretations of Aboriginal Australian landscape burning practices and their relevance to contemporary ecosystem management. To begin with, there is an abundance of scientific literature on Aboriginal Australian fire regimes, probably due to the relatively undisturbed continuance of traditional burning technology in the northern regions of the country. On the whole, such scientific narratives of burning strongly construct the aesthetic appreciation of fire as a utilitarian phenomenon to be controlled or an unnatural occurrence to be avoided. On an ecosystem management level, the absence of Aboriginal landscape burning in remote areas has resulted in intense seasonal fires. Growing scientific evidence reveals that the present pattern of landscape burning is deleterious to the ecological health of large areas of the Australian savanna, causing species extinction, devastation of fragile habitats such as rainforests, and degradation of ecosystem processes such as carbon sequestering and soil fertility (Bowman and Vigilante 2001). In the absence of sustained human intervention, periodic large and high intensity wildfires can cause broad-scale change in the condition of the vegetation and its associated biota.

An example of Australian Indigenous burning and ecosystem management comes from the northern Kimberley region, located in the far northwest of Western Australia. Like the rest of the Australian monsoon tropics, the vegetation of the area is tall-grass Eucalyptus savanna (Bowman and Vigilante 2001). If this terrain goes without burning for a few years, spinefex grasses (*Triodia* spp.) mature into large hummocks and become resinous. Burning the spinefex at an early stage is vital to encourage the emergence of fresh new shoots instead of the development of clustered old plants, which yield a resin that makes kangaroos sick. In the absence of periodic burning of the savanna, the establishment of other tenacious weeds, such as Gamba grass, can cause the fires to achieve an intensity that destroys important native trees.

Scientific research into fire systems indicates that a shift in the spatial scale of fires has resulted in the loss of fire sensitive flora and the preponderance of fire-promoting grasses (Yibarbuk et al. 2001, 336-337). Lowland savannas, mostly dominated by native annual sorghum, annually accrue significant fuel loads capable of sending extensive fires into sensitive vegetation areas such as those of escarpments. Such conflagrations accumulate enough fuel one to three years after the last fire, a period much shorter than is required for regeneration of many obligate seeders that need complete protection from fire from seedling establishment to achievement of reproductive maturity.

Dukaladjarranj, a clan estate in north-central Arnhem Land in the seasonal tropics of northern Australia where Aboriginal occupancy has been close to continuous, demonstrates the connection between firing the landscape, traditional management objectives, and ecosystem/floristic composition. Annual sorghum, the major contributor to dangerous fuel load levels elsewhere in the savanna, is all but absent (Yibarbuk et al. 2001, 337). Furthermore, a high frequency of healthy stands of Cypress pine (*Callitris intratropica*) in areas that had been recently fired indicate the beneficial effects of patchy, low intensity fire. Sustained Aboriginal management and fire use in Dukaladjarranj has maintained a regime of contained, high incidence, and low intensity disturbance, to which the modern biota are exceedingly resilient.

In western Arnhem Land, traditional fire management is also effective in managing the landscape (Russell-Smith n.d.). Aboriginal burning practices here, like those of other clans in remote parts of northwestern Australia, are related to such considerations as the habitat types and resources involved and the faunal relationships between habitats (Lewis 1989). Regular burning of floodplains kept fuel loads low, especially where slow burning peat fires or *mulurr* would burn until suppressed by rains (Russell-Smith n.d.). *Mulurr* was used to control thick stands of native hymenachne (*Hymenachne acutigluma*) and to promote the growth of valued food plants such as water chestnut (*Eliocharis* spp.). In fact, the regular burning of floodplain grasses might have restrained natural rainforest development.

The south and central areas of the west coast of Australia provide further instances of traditional burning technology and ecosystem management, particularly related to the concentrated use of "fixed patch" root crop resources (Hallam 1989). Burning of the country provided access to some resources such as yam (*Dioscorea hastifolia*) that were carefully preserved in unburned areas and encouraged others like zamia (*Zamia* spp.) and pasture grasses (Hallam 1989, 143). Aboriginal burning established a mosaic pattern of open country and thick brush, creating zones of easy travel, and opening up access to the western coastal plain and eastern woodland margin. The selective use of fire, in conjunction with the knowledge of how and when to employ burning and where to limit it, made possible an elaborate subsistence system of burned and preserved areas.

In Tasmania, 240 kilometers from mainland Australia, some of the rainforest and alpine environments of south-west Tasmania's World Heritage Area have been in decline due to an increase in the incidence of fires. Furthermore, some of the buttongrass (*Gymnoschoenus sphaerocephalus*) moorland hummocks that depend on the region are threatened because of a decreased incidence and size of burns (Marsden-Smedley 2000, 195). Eucalypt forest (*Eucalyptus* spp.), tea-tree scrub (*Acacia* spp., *Banksia* spp., *Leptospermum* spp., and *Metaleuca* spp.) and buttongrass moorland in south-west Tasmania are products of a fire regime consistent enough to prevent the successional process that terminates in rainforest. The regime utilized by Indigenous people to maintain these ecosystems in south-west Tasmania was one of frequent and probably relatively low-intensity fires (Marsden-Smedley 2000, 196).

In the Victoria River valley of the Northern Territory, Yarralin land managers are guided in their interactions with country through the notion of *punya* (Rose 1992). *Punya* refers to human actions which make everything "come up new and fresh," make "the country happy," and make "the country clean and good" (Rose 1992, 100-101). *Punya* also points to nurturing the land to allow for good hunting conditions, abundant kangaroos and wallabies, and the seasonal regeneration of flora and fauna. A vital aspect of caring for country in the Yarralin view is traditional burning. As Deborah Bird Rose comments, the firing of the landscape is part of a larger cultural narrative: "This is a cyclical process in which the knowledge and care which humans put into the system, including the former deposition of bones, form an essential part of human survival in the system, making people, other species, and country *punya*" (Rose 1992, 101). Moreover, Rose cites an unidentified Aboriginal correspondent in *Nourishing Terrains* as correlating devastating bushfires with lack of frequent low-intensity burning: "Big fires come when that country is sick from nobody looking after with proper burning" (Rose 1996, 66). Hence, the narratives spoken by Aboriginal peoples themselves, suggest the centrality of firing practices to cultural and spiritual well-being.

The Role of Journalistic Narratives in Popular Appreciation of Fire and Ice

As a narrative form linked to the aesthetic appreciation of the environment, a *journalistic* account will encompass, for the purposes of this discussion, the diaries of explorers, adventurers, and homesteaders, as well as media representations of the natural world. In terms of the latter, a corpus of material has emerged during the last thirty years, spurred by public fascination with mountaineering and public horror over wildfires out of control, particularly in Australia. Journalistic accounts—either in the form of historical representations in the diaries of an explorer or as contemporary works by environmental writers—are distinct forms of narrative, characteristically evolving out of an individual person's contemplations. Journalistic depictions of the natural world are crucial to consider in relation to environmental aesthetics because they intrinsically weave together subjective impressions, universalizing discourses, science-based analysis of the history of an environment or place, and the social and cultural implications of events such as wildfires. To state it differently, a personal story can become part of natural history. For instance, consider the following passage from the journal of Alfred H. Brooks, who in 1902 became one of the first Europeans to ascend Denali:

> Climbing the bluff above our camp, I overlooked the upper part of the valley, spread before me like a broad amphitheatre, its sides formed by the slopes of the mountain and its spurs. Here and there glistened in the sun the white surfaces of glaciers which found their way down from the peaks above. The great mountain rose 17,000 feet above our camp, apparently almost sheer from the flat valley floor [...] I found my route blocked by a smooth expanse of ice. With the aid of my geologic pick I managed to cut steps in the slippery surface, and thus climbed a hundred feet higher; then the angle of slope became steeper, and as the ridge on which the glacier lay fell off at the sides in sheer cliffs, a slip would have been fatal. Convinced at length that it would be utterly fool-hardy, alone as I was, to attempt to reach the shoulder for which I was headed, at 7,500 feet I turned and cautiously retraced my steps, to find the descent to bare ground more perilous than the ascent. (Beckey 1993, 61)

The abrupt contrast between the flat valley floor and the sheer verticality of the mountain is expressed structurally through the image of an amphitheater, which further suggests the imminent drama of ascent. The "dome-shaped summit" and "upper slopes [...] white with snow" make appreciable the aesthetic properties of the mountain. A reader can take on the perspective of the historical subject, standing at the base of Denali, head hyper-extended in order to take in the height of the massive mountain. The immense landscape overwhelms the explorer; expanses of ice and loose talus complicate walking; its features seem to engulf the human. Additional efforts would be "fool-hardy." Thus, the reader gains an impression of Denali that combines detailed representation of aesthetic qualities with overtly personal expressions of awe and fear.

Unlike the technical language of science and the culturally and linguistically specific narratives of Indigenous peoples, the prosody of journalistic accounts blends individual impression with perceivable aesthetic qualities and often some reference to scientific knowledge. Yi-Fu Tuan appraises this distinguishing aspect of journalistic accounts of nature in the diaries of arctic and desert explorers in which the confrontation with death in extreme climates tends to permeate the prose. For instance, polar explorer Fridtjof Nansen couples the beauty and splendor of Arctic ice with death: "Unseen and untrodden under their spotless mantle of ice the rigid polar regions slept the profound sleep of death from the earliest dawn of time" (quoted in Tuan 1993, 151). By contrast, Richard Byrd's language was more optimistic but occasionally drifted into a melancholy tone: "A funereal gloom hangs in the twilight sky. This is the period between life and death. This is the way the world will look to the last man when it dies" (quoted in Tuan 1993, 153). Further quoted by Tuan, Byrd also aligned the pulsing of the aurora with classical music: "So perfectly did the music seem to blend with what was happening in the sky" (1993, 153). According to these journalistic accounts, the boundary between the inner world of sensation, emotion and memory, and the external domain of aesthetic objects becomes indistinct in extreme environments.

The effectiveness of journalistic material, as defined here, lies in its capacity to stimulate an aesthetic response to the natural world. Aesthetic (human) subjects are drawn into an otherwise static scene/seen through tangible language that expresses emotional and psychological states identifiable to the reader. In this context, Heyd's linkage of environmental perception to narrative is again germane (Heyd 2001). The detailed treatment of a natural object or scene laden with individual impressions invites us to consider the aesthetic properties communicated by the narrative. We select (either consciously or not) certain properties to inform our own responses to mediated nature. Saito's moral imperative to appreciate nature on its own terms, without distortion or falsification, guides us in accepting certain aesthetic elements within the narrative as compelling, believable, or appropriate. For example, aesthetic responses to arctic environments, as described in Nansen's journal, can be infused with the same overtones of death and austerity. Whether we reflect the mood of foreboding in our aesthetic reaction to ice or rock depends on the level of our intention to represent the environment fully and appropriately. Furthermore, we can consider Denali metaphorically as an amphitheatre in order to appropriate the juxtaposition of horizontal and vertical features for our aesthetic response.

Additionally, we could choose to see burnt country as cared for or as devastated, depending on the acuity of our sensory intake and severity of our memories. I emphasize choices (i.e. consciousness) here, with regard to all three narratives discussed so far, because a critical aspect of aesthetic appreciation of the natural world relates to what we decide to use as the content of aesthetic meaning-making. A practice of aesthetic hybridization between features of each narrative facilitates a critically pluralistic environmental aesthetic.

Returning to the Aesthetics of Burning: Examples from North America and Africa

As the previous examples from Aboriginal Australia suggest, the use of fire to maintain optimal habitat conditions and stimulate vital plant and animal resources has been a common practice in many Indigenous societies (Turner 1999). Although the reasons for firing landscapes vary broadly—e.g. driving game, improving visibility, promoting ease of travel, driving away noxious pests like mosquitoes, increasing the supply of food plants, maintaining habitat for game species, fire-proofing villages, and as defensive or aggressive measures in warfare—a common denominator of burning practices across cultures is the role of fire in sustaining the dynamic equilibrium of ecosystem composition (Day 1953). As part of complex and ecologically specific systems of environmental management, traditional burning is a strategy for maximizing subsistence resources. The focus of this section is to examine contemporary and historical documentation of traditional burning regimes from eastern and western North America and Africa with reference to the consequences of each strategy on ecosystem structure and to address the diverse stories of burning along the way. As these examples will show, the conscientious selection of certain plant species, the suppression of less important food species, and the subsequent creation of favored habitat for mammals resulted in distinct environments administered on a regular basis by traditional land managers.

The Indigenous people of the northeastern United States routinely burned the forests and grasslands for many reasons, including to encourage plant and animal resources and maintain a clear forest understory (Day 1953, 334). Journals from the European settlers to the northeastern areas of North America are replete with descriptions of the manicured (i.e. picturesque, a category of aesthetics) appearance of forests. These accounts commonly cite traditional burning as the reason for the absence of forest floor cover. The open quality of the forests impressed early settlers and evoked images of the woodland parks of Europe. As cited by Day, Lindstrom recorded the profusion of tall grass and the trees which "stand far apart, as if they were planted," concluding that the burning of the dry grass for the spring hunt maintained such open conditions (Day 1953, 334). Also cited in Day, Morton attributed the tended quality of the woodlands to the "firing of the country" (1953, 335) and Johnson observed that the woods were made "thin of timber in many places, like our parks in England" through frequent burning (1953, 334). The colonial narratives describing the landscapes, created through the burning regimes of indigenous North American peoples, share commonalities with those in Australia around the same time, as previously discussed in this chapter.

Grasslands and open oak forests dominated the plant communities of present-day New York and New Jersey with the distinct "oak openings" characterizing the vegetation pattern (Day 1953, 338). The journals of settlers emphasize the openness of the upland woods, detailing oak, chestnut, and hickory communities on the slopes, white pine and hemlock in the swamps, and bushy plains and blueberry barrens on the plateaus and hilltops (Day 1953, 337). Moreover, Russell argues that good timber could only be found in the lowlands since periodic burning of the uplands prevented big trees from maturing (Russell 1983). After traditional fire management ceased, pine species began to encroach upon the oak-chestnut communities, and the oak openings diminished as the growth of shrubs went unimpeded (Day 1953, 338).

In the woodlands of eastern North America, the frequency of fires affected the forest composition, according to the intensity of the burn and the vegetation structure before the fire (Russell 1983, 86). Recent studies of the effects of fire on oak regeneration suggest that occasional light fire improves oak re-growth for some species while more frequent fires destroy trees. Thus, less frequent burns of low intensity probably contributed to the oak communities observed from southern Maine to Virginia in pre-colonial historical documents. Additionally, grass seems to have been the ground cover where fires occurred, in present-day New York and New Jersey, indicating that selected places were burned at close intervals while others were left untouched.

In western North America, fire regimes were also utilized to manage ecosystem structures and to enhance the availability of food species. Turner describes that, apart from the increase in plant productivity, Indigenous peoples of western North America burned the landscape to enhance forage for deer and other game (Turner 1999, 200). In all, at least nineteen species of plants, "including eleven fruiting shrubs, one herbaceous fruit (strawberry) and seven herbaceous 'edible root' species, have been identified by various sources as having their production enhanced by periodic burning" (Turner 1999, 188). All of the species indicated by Aboriginal peoples of western North America to benefit from periodic landscape burning have the capability to regenerate from buried rhizomes or subterranean storage organs and are successional plants (Turner 1999, 201). However, as bushes encroached on the hills after the cessation of yearly burning, ripe berries vanished and roots collectively called "potatoes" (*Erythronium, Lilium columbianum,* and *Claytonia lanceolata*) disappeared (Turner 1999, 189).

Indigenous people's firing of the land promoted the growth and distribution of preferred resource species. For instance, the Kalapuya of the Willamette Valley in Oregon burned hazelnut (*Corylus cornuta*), an early fire follower that flourishes on charred sites (Turner 1999, 192). Burning the mountainsides promoted the growth of onions (*Allium cernuum*), raspberries (*Rubus idaeus*), blackcaps (*Rubus leucodermis*), and probably huckleberries (*Vaccinium membranaceum*) (199). There is persuasive evidence for traditional landscape burning of blue camas (*Camassia quamash* and *C. leichtinii*) prairies in western Washington and large-scale burning by Aboriginal peoples on Whidbey Island, Washington (195). Alaska blueberries (*Vaccinium alaskaense*), red huckleberries (*V. parvifolium*), and salal berries (*Gaultheria shallon*) also respond especially well after an area has been fired (198). Additionally, evidence definitively indicates that burning influenced the range of certain species. The Tsolum River area of British Columbia is the northernmost location for garry oak (*Quercus garryana*); unlike other garry oak sites, the area falls within the Western Hemlock Biogeoclimatic Zone. Therefore, Page asserts that these oaklands were maintained in the past by periodic burning (197).

The ongoing need for young plant growth provoked Indigenous western North Americans to carry out periodic burning of sites to enhance the availability of basketry materials. Long straight shoots with no lateral branches that emerge after a section has been burned over were preferred for the manufacturing of many products: "The most favored shrubs, grasses, ferns, and sedges for basketry, as well as the preferred herbaceous plants for edible corms, bulbs, and tubers, all evolved and thrived in a context of periodic disturbances that included flooding, rodent activity, fire, and herbivory" (Anderson 1993, 156). For example, the Karok and Wiyot burned to make hazel (*Corylus cornuta*) and willows (*Salix* spp.) yield superior materials for manufacturing baskets (1993, 163). Also, deer grass (*Muhlenbergia rigens*) was managed with fire to augment flower stalk yield and to bolster the dimensions of the culms (1993, 174).

This chapter began by citing the image of the railroad line delineating forest to one side and prairie to the other in northern Alberta. An analogous scenario is found in the film *Second Nature*, where in a village near the Prefecture of Kissidougou in Guinea, the researchers stand at the abrupt edge between the fired savanna and the maturing forest (Maughan 1996). On a broader geographical level, the forest-savannah transition zone of Guinea is the interface between the equatorial rainforest to the north and the arid lands to the south. The people of Kissidougou employ fire to manage this dynamic zone and, in doing so, increase the fertility of the soil, the abundance of plant resources, and the safety and comfort of their villages. For the Dene, the winter collection of rabbits is augmented by spring burning; by inducing early growth or early succession of plants during the spring burn, the plants achieve more robust growth to fortify them through the winter and make available plant nutrients to foragers. Similarly, for the people of Kissidougou, a regime of burning constitutes a seasonal resource management technique. Burning is strategic in both cultures. The ecological premises underlying Kissidougou decisions to initiate a burn are as equally complex for the Dene. For example, the Kissidougou firers balance between white grass and long grass, cognizant of how a predominantly long grass environment will transition to forest whereas a white grass savannah will not. Thus, the choice to facilitate forest growth functions with the management of savannah grasses.

The firing strategies of both cultures maintained a dynamic balance between grassland and forest. The major distinction, however, between the northern Dene and Kissidougou systems of burning is that the Dene strategy aimed to maintain the grassland prairies of northern Alberta's boreal forest while the firing system of the people of Kissidougou encourages productive forest rings around the villages. The cessation of Dene burning has resulted in the encroachment of forest, obscuring once long-spanning views over the rivers of the land, as related by the elders. By contrast, the modern increase in Kissidougou burning, as part of a comprehensive ecosystem management scheme, has resulted in the deliberate encircling of villages by forests where savannah dominated even fifty years prior.

Other differences between the systems include the plant species involved; the Dene once actively managed an aspen, birch, and willow parkland whereas the Kissidougou forests are mostly red oil palms, policion, coffee trees, kola, and silk cotton trees. And while there are clear ecological implications in their fire management practices, there are also aesthetic dimensions embedded in these narratives of burning.

Narratives of Fire and Ice in Dialogue: Representing the Natural World in Critically Pluralistic Terms

This chapter has examined a cross-section of narratives that can facilitate our aesthetic regard for the nature of fire and ice, savannah and mountain. Carlson's model claims the narrative of science as the most appropriate one. However, we have seen, using Denali and the burning practices of Indigenous peoples across the world as examples, that appropriate (i.e. broad-based, culturally diverse and ontologically congruent) aesthetic responses can be derived from Indigenous and journalistic accounts as equally and effectively as scientific discourses. This is particularly evidenced by the discussion of firing practices in different cultures as observed by "outsiders" and expressed by "insiders." Whether we give credence to the subduction of tectonic plates or the crystallization of enormous waves agitated by the treachery of Raven should not matter. The aesthetic experience proffered by both—the symphony of abrasion, slipping, scuffing, and sculpting arranged by astounding geological forces adhering to a nearly incomprehensible standard of time; the elegant transfer of energy and form from the wave of a lake to the crest of the mountain; an exchange roused by the misdeeds of a personified raven; or an aesthetic sense for country as clean and cared for—convey the distinctive essence of a landscape, prompting a sensible aesthetic response to these places, hence revealing them in written terms, incrementally for what they are. Furthermore, journalistic accounts, though inherently combining subjective and objective impressions, present a story of the natural world through which we can navigate aesthetic values. Whether our aesthetic appreciation is a misguided or accurate one depends on the sources of "data" with which we engage. Our appreciation of nature can become critically aesthetic without turning toward monolithic notions of truth and essentialist forms of science.

I have argued that diverse narrative forms—science, Indigenous cosmologies and ecological knowledge, and journalistic accounts—can provide equally substantial representation of aesthetic qualities. The aesthetic (human) subject might be satisfied to consider the harmoniously separate existence of such narratives, each artfully employed by the groups (scientists, Indigenous groups, and journalists) who promulgate them. Let the geologists have their fill of tectonics and the tediously slow movement of landmass; Indigenous peoples their fascination for Raven, the spontaneous generation of landscape protrusions in their Dreaming stories, or their ecocultural narratives of fire; the explorer his personal version sketched from a single instance of exposure to the vicissitudes of the mountain and filtered through written language in historical documents. The "two cultures" gulf lessens when we invite narrative forms to intermingle; and to verify the degree to which the other makes appreciable the real aesthetic properties of a specific natural feature embedded in a distinct ecology and cultural milieu. I assert that this "cross-checking" of aesthetic narratives offers a sensitive yet critical reflection of the natural world as an ecocultural creation. Holmes Rolston states: "Metaphysical fancy has to be checked by a pragmatic functioning, and this includes an operational aesthetic with some successful reference to what is there at one's location" (Rolston 1995, 383). Unlike Rolston who implies a hierarchy, I maintain that each narrative should bear weight in a critically pluralistic environmental aesthetic. Accordingly, scientific narrative need not eclipse the cosmologies and practices of Indigenous peoples. Stories of consensus—those shared by groups of people and generations as cultural works (science and mythology)—verified by their antiquity or the "test of time," should not supersede the obviously subjective representations of the natural world in past and present journalistic accounts.

Returning to a critically pluralistic aesthetic of Denali as an example, the assertions of geology—the uplift of plates and the eroding action of glaciers—would be understood only in reference to the eye-witness stories of explorers. Conversely, the explorers' accounts could bear witness to the talus jumble and the glacial sheets, confirmed by geology's rationalization of the same geological landform. Furthermore, Indigenous knowledge would confirm the unique perceptual features of the mountain summit, the semblance of a frozen wave, which in turn would be confirmed by first-hand personal exposition and scientific explanation of Denali's glaciation and high-altitude wind patterns. This process, which I roughly call "cross-checking," involves discerning aesthetic realities, passing multiple accounts through the sieve of human perceptual and cultural diversity to separate the most developed, established, and nuanced narratives from mere idiosyncratic flights of fancy that do not represent nature in its own terms. Particularly with the advent of digital media during the last twenty years, individuals access multiple environmental narratives sequentially or on a one-by-one basis according to their interests, locations, or practices at any given time. Hence, there is a multitude of ways people access and interpret narratives; and thus a need for a philosophy of environmental narrative heterogeneity to guide individuals in their knowledge acquisition and value formation.

Concerns endemic to "the story"—for instance, multiple acceptable interpretations; the anthropocentric humanizing of nature; and the representation of the natural world through a cultural bias—can be addressed when each narrative is critically examined against another, rather than through the internal logic of a "discipline," as with scientific narratives in particular. Through cross-checking, geological narratives would help to illuminate cosmological stories, which might reciprocally illuminate journalistic accounts in order to distill appreciable aesthetic qualities into an integrated whole. As such, the wave-like patterns of snow atop Denali are identifiable in all three narratives. Aesthetic exaggerations—those isolated and possibly distorted representations of nature (Denali's slopes flanked in gold sheets!)—would be immediately identified as anomalous and possibly dispelled or further investigated. In this manner, our fanciful projections upon the natural world would be kept in check without dismissing imagination—that significant human dimension of experience of the environment, often marginalized in light of Cartesian objectivity or Kantian disinterestedness. Where multiple stories mutually illuminate one another, at such an intersection, a narrative environmental aesthetic emerges to represent nature to the most inclusive degree possible by integrating elements of seemingly disparate stories: scientific, Indigenous, and journalistic as the three selected for this chapter.

Conclusion: Toward Critical Pluralism through Narratives

An environmental aesthetic philosophy based upon multiple narratives of nature is an embryonic possibility, in part, because narrative is regarded as a fringe component of environmental aesthetics. Cross-checking constrains our interpretations of the stories of nature, enhancing and grounding our appreciation by filtering away narratives that are probably not representative of nature and therefore not appropriate. Science, Indigenous cosmologies and practices, and journalistic accounts constitute a larger story, a dynamic and integrated human history intimately connected to a planetary natural history. In this chapter, I have argued that stories are part of an holistic environmental aesthetic that gathers human and non-human senses, emotions, moods, and imaginative faculties. Indeed, a narrative natural aesthetic can be one of critical pluralism. Emily Brady writes that "critical pluralism sits between critical monism and 'anything goes,' the subjective approach of some post-modern positions. It argues for a set of interpretations that are deemed acceptable but which are not determined according to being true or false [...] An interpretation must be acceptable, it cannot be outlandish, irrelevant or the whim of one person" (Brady 2003, Chapter 3). In terms of the general moral derived from the case studies of snow and fire, I align with Brady's "critical pluralist" approach which seems to identify the acceptability of aesthetic interpretations rather than determine the truth or falsehood amongst them. Hence, the environmental aesthetic syncretically derived from scientific, Indigenous, and journalistic perspectives is not based on polarized value judgments—which could demarcate and set in opposition those perspectives—but rather on their points of intersection. The mechanism of "cross-referencing," which seeks commonalities and nodes of connection, facilitates the assessment of acceptability along a continuum from "outlandish" and "irrelevant" to appropriate for certain environments and acceptable to the well-being of the human and more-than-human life residing there.

Through critical pluralism, "cognitive" aesthetic theories, such as Carlson's natural environmental model, and "non-cognitive" approaches, based in imagination and mythologies, dynamically co-exist, amplifying one another to represent nature on its own terms. An integrated environmental aesthetic benefits from the complementarities of different narratives and aesthetic modes. Indeed, "non-cognitive" modes, represented by human imagination, moods, and emotions, provide important components deficient in scientific narrative approaches yet crucially important to human appreciation of nature. Indeed, as feminist philosophical work argues, the non-cognitive (subjective, imaginative, intuitive, emotional) and cognitive (objective, rational, logical, deductive) binary is an assumption that plagues much ethical theory. Rather than conceptualizing subjective, imaginative, and poetic interpretations of reality as either true or false (and thus reverting to traditionally "cognitive" criteria), a critically pluralistic aesthetic reflects Brady's notion of acceptability as the interpenetration of the "cognitive" and "non-cognitive" in any form of aesthetic experience of the natural world (Longino 1997). Simply put, one mode of experiencing and/or representing nature narratively should not come at the expense of another. A critically pluralistic environmental aesthetic would draw from all of our senses, our intellect, and our personal proclivities.

As David Sobel has observed, "environmental empathy" develops from experience and imagination as "a feeling for other creatures that can develop into a willingness to care for other creatures" (Sobel 2008, 30). Sobel argues that empathy establishes the groundwork for learning ecological science later in life. As I have tried to demonstrate through the stories of Denali and those of Indigenous firing practices, a productive and critical tension between all modes of narrative ensures that nature is appreciated and valued appropriately. The articulation of narrative heterogeneity in the formation of a critically pluralistic environmental aesthetic—where diverse expressions of nature mutually illuminate one another—should be the subject of further considerations. This is especially urgent because, as I sit here writing in January 2013, a constellation of blazes—amplified by temperatures above 100 degree Fahrenheit, powerful winds, and low levels of winter rain—threaten many parts of Tasmania, New South Wales, and Victoria in Australia. Undoubtedly in relation to these fires, new narrative forms will surface in the media—the complaints of residents against emergency services providers; the stories of locals who have lost their homes; memories of these catastrophes in relation to catastrophes from other years; the reminders of scientists who attribute more intense fires to climate change; and the wisdom of Aboriginal peoples who maintained small-scale fire regimes in Australia for over 40,000 years. One of the challenges of our age is to encompass these diverse narratives—ever vigilant for their aesthetic, sensory, and bodily implications for human and more-than-human well-being—into integrated forms of knowing *about*, being *with*, and co-existing *between* and *within*.

Chapter 3: Plant Narratives and the Senses: Thoreau's Approach to the Botanical

Introduction

Take their cold seed and set it in the mind,
and its slow root will lengthen deep and deep
till, following, you cling on the last ledge
over the unthinkable, unfathomed edge
beyond which man remembers only sleep.

—Judith Wright "The Cycads" (Wright 1994, 39–40, 17–21)

Science, history, poetry, mythology and personal experience are often thought to contradict one another and are regarded, therefore, as separate forms of knowledge-making. Like leaves of a tree, however, the botanical works of Henry David Thoreau bring together the diverse stories that give meaning to the natural world, as described in the previous chapter. Drawing from the concept of *multiple narrative streams* as a method of writing about nature and doing natural history, as inspired by Thoreau, Chapter 3 will explore different historical accounts of the flora of the South-West of Western Australia. Botanical sciences, Indigenous spiritualities, nature poetries such as Judith Wright's and colonial histories offer disparate though complementary perspectives. The intertwining of narrative streams ensures the perpetuity of non-scientific stories and the potential for cross-pollination between disciplines and multiple ways of knowing flora.

Chapter 3 suggests that, as an approach to writing about plants, it is crucial to consider how stories—including poetic and scientific—augment and amplify each other, rather than promulgate divisions between human culture and botanical nature. Importantly for this discussion, the narratives of plants contain distinctive aesthetic modes and values. Through verse instead of science, Judith Wright evokes the primordial character of the cycads, survivors of the age of the dinosaurs and older than the human species. The final line of the poem "beyond which man remembers only sleep" intimates the qualities of adaptation and co-evolution that may distinguish the cycads from recently introduced taxa. Convergent Aboriginal, poetic and scientific stories are etched in the slow-growing cycad fibers. A polyvocal account of the cycads weaves together technical views, cultural histories, Aboriginal understandings and multi-sensorial aesthetic experiences. The greater cohesion of narratives (informed by the "cross-checking" approach outlined in Chapter 2) results in a productive confluence for writers of botanical history, environmental issues and philosophical aesthetics.

Through Thoreau's sense-rich writings, the approach of multiple narrative streams will be outlined in this chapter as a basis for understanding the diverse narratives of South-West Australian flora. Indeed, this approach could be applied to any account of natural history, from animals and birds to rocks and algae. Barker (2008, 483) defines narrative as "a sequential account or purported record of events ordered across time into a plot. The concept of narrative refers to the form, pattern or structure by which stories are constructed and told." Although commonly restricted to literary theory, *narrative* will be used synonymously with the terms *story*, *account* and *history* in reference to plants. Nyoongar histories, early settler accounts, post-colonial narratives, poetic interpretations and experiential impressions broaden the dominant narrative of science towards storied streams that diverge and converge dynamically. Through narratives, disciplinary accounts that historicize plants, such as those offered by natural history, are networked together with the diverse narrative forms explored in Chapter 2. An emphasis on the commingling of knowledge streams minimizes the potential erasure of certain accounts by dominant, techno-scientific narratives (for an example of commingling, see Hopper 2010).

Thoreau's Multiple Narratives of Nature

Throughout his sustained criticism of science and empiricism, Thoreau's botanical understandings were continually augmented by his own experiential involvement with flora over his lifetime. For instance, Thoreau's empiricism led to the development of a novel theory of seed dispersion. His ideas resulted in an intellectual space of convergence between scientific, sensuous and Indigenous modes of valuing flora (Chapter 2). His works exemplify the seamless integration of narrative streams towards multi-faceted and epistemologically intricate accounts of plants. Thoreau proffers an approach to flora in *Faith in a Seed* (1993) and *Wild Fruits* (2000), which mark his growing intrigue with field botany and forest ecology beginning in the early 1850s. Although his earlier pièce de résistance *Walden* alludes to local species, an interest in flora consumed his later writings, as this journal entry from 1856 indicates:

> I soon found myself observing when plants first blossomed and leafed, and I followed it up early and late, far and near, several years in succession, running to different sides of the town and into the neighboring towns, often between twenty or thirty miles in a day. (Thoreau 1962, 158)

Thoreau planned to assemble his observations of flowering and leafing into a "Kalendar," modelled after John Evelyn's 1664 *Kalendarium Hortense* or *Gardener's Almanac* (Dean 2000). The project would be a comprehensive phenology of an archetypal year, setting out all the events of Concord natural history. The work would strive for comprehensiveness and eclecticism spanning scientific botany, Native American ethnobotany, classic Greek and Roman philosophies and Thoreau's first-hand experiences.

Although Thoreau passed away before his ambitions were realized, the posthumous *Wild Fruits* gives a sense of his methodology, particularly his use of narrative multiplicity as an overarching framework. In his exposition of the strawberry, Thoreau (2000, 10-17) begins with verses composed by the sixteenth-century poet Thomas Tusser and a description by the herbalist John Gerard, writing in the pre-Linnaean 1500s. Thoreau (2000, 11) reflects upon his observations of wild strawberries, stating "by the thirtieth of May I notice the green fruit." In this evocative passage, he expresses the fragrance of wild strawberries as a quality that evades the visible prominence of its flowers. Strawberries emit:

> An indescribably sweet fragrance, which I cannot trace to any particular source. It is, perchance, that sweet scent of the earth of which the ancients speak. Though I have not detected the flower that emits it, this appears to be its fruit. It is natural that the first fruit which the earth bears should emit and be, as it were, a concentration and embodiment of that vernal fragrance with which the air has lately teemed. Strawberries are the manna found, ere long, where that fragrance has been. Are not the juices of each fruit distilled from the air? (Thoreau, 2000, 12)

Thoreau also conveys the impressions of explorers to North America, including eighteenth-century Englishman Samuel Hearne's observations of the strawberry in Indigenous North American cultures. *Oteagh-minick* in the language of First Nations Canadians of the Churchill River signifies the resemblance of the fruit to a heart, while other names for strawberries, as *Oteimeena* in Cree and *O-da-e-min* in Chippeway, present a visual signature linking the human heart to the shape of the fruit (Thoreau 2000, 15).

Thoreau studied the writings of North American explorers and early ethnographers, citing the American theologian Roger Williams's landmark book on Native American linguistics, *A Key into the Language of America* (1643). On the strawberry, Williams (quoted in Thoreau 2000, 16) reveals ethnobotanical interest and what would be called by modern ethnographers a form of participant observation: "The Indians bruise them in a mortar, and mix them with meal, and make strawberry bread [...] having no other food for many days." Further in his exposition, Thoreau draws from the records of naturalists, such as the missionary George Loskiel, to determine changes to the distribution of strawberries in the eastern United States. Through this vignette, Thoreau exemplifies the use of narrative streams by juxtaposing historical texts, ethnographic accounts, multi-sensorial interaction and astute personal observations. This syncretism augments scientific conjecture, an approach he perfects in the posthumous *Faith in a Seed*.

Echoing his approach to strawberries, the essay "Wild Apples" (Thoreau 1862/2010) synthesizes field science, references to classical writers, and the author's corporeal experiences in the field. Thoreau's reading on wild apples is wide-ranging, comprising allusions to Tacitus, Palladius, and Pliny. Multi-sensorial immediacy mingles with the voices of previous writers toward a cultural botany of the wild apple. Thoreau (1862/2010, 25) qualifies some varieties as "acrid and puckery, genuine verjuice," while a particular tree in Concord produces "a peculiarly bitter tang, not perceived till it is three-quarters tasted. It remains on the tongue. As you eat it, it smells exactly like a squash-bug" (Thoreau 1862/2010, 27). The essay celebrates the sensuousness of apples in an age of the increasingly homogenized sizes, shapes, tastes, colors, and smells of fruit. Thoreau (1862/2010, 28) beseeches the reader to "let your condiments be in the condition of your senses. To appreciate the flavor of these wild apples requires vigorous and healthy senses, papillae firm and erect on the tongue and palate." Although a sensuous naturalist, he relied on emerging botanical science. In the section "The Naming of Them," Thoreau (1862/2010, 30) affirms the advantages of standard nomenclature: "I find myself compelled, after all, to give the Latin names of some for the benefit of those who live where English is not spoken,—for they are likely to have a world-wide reputation." As suggested by these excerpts, Thoreau engaged different streams of knowledge, guided by his intense curiosity, without attempting to rationalize how they might lock together epistemologically (see Chapter 2). Thoreau's writing demonstrates the confluence of knowledge-making approaches involving cross-fertilization between personal, poetic, multi-sensorial, historic, Indigenous, and scientific ways of knowing.

The Narrative Streams of Zamia Palm

A case study will illustrate the concept of narrative streams applied to the study of the flora of the South-West of WA—a region of remarkable plant diversity alluded to in the discussion of the Nyoongar seasons in Chapter 1 and outlined more fully in the Preface. The zamia palm (*Macrozamia riedlei*) is a member of the Zamiaceae family of cycads distributed throughout Australia, Africa, and warm temperate areas of North and South America. First classified by Charles Gardner, *M. riedlei* is endemic to the lower South-West corner of the South-West region, from Hutt River near Perth to Albany. The plant contains macrozamin—a toxin found in most cycads that is responsible for zamia staggers, a fatal affliction of the nervous systems of animals (Carr and Carr 1981, 18). In the South-West, zamia palms are thought to exist only in iron-rich lateritic soils and as understory plants in jarrah forests. However, other botanists observe that the species is common in all soil types throughout Perth (Marchant et al. 1987, 57). Zamia demonstrates a short trunk, about three meters high, rigid fronds one to two meters in length, and broadly cylindrical or ovoid seeds that are reddish brown and fleshy when ripe (Marchant, et al. 1987). Reflecting the condensed structure of most taxonomic descriptions, Paczkowska and Chapman (2000) offer this morphological snapshot:

> **Macrozamia riedlei** (Gaudich.) C.A.Gardner ZAMIA
> Cycad, 0.5-3 m high; small, usually trunkless; leaves few, glossy,
> flat or openly keeled, narrow leaflets; short cones. Fertile
> plants recorded Sep-Oct. Lateritic soils, jarrah forests.
> Distribution: SW:ESP, GS, JF, SWA, WAR. (27)

When the Dutch explorer William de Vlamingh landed in December 1696 on Rottnest Island, west of the mouth of the Swan River, his party found zamia nuts.

Vlamingh reported the initial palatability of the roasted fruits, likening them to "Dutch broad beans, or, when ripe, like hazelnuts," but three hours after consuming them, his crew "began to vomit so violently that there was hardly any distinction between death and us" (Vlamingh 1985, 155). In January 1802, with comparable indiscretion provoked by a deadly mix of desperate hunger and genuine curiosity, members of the Flinders expedition at Lucky Bay near Esperance were drawn to eat zamia fruits, but with equally disastrous consequences: "A party of gentlemen were upon the top, eating a fruit not unlike green walnuts in appearance [...] Mr. Thistle and some others who had eaten liberally were taken sick and remained unwell all the day afterwards" (Flinders 1814, 80). Beaton (quoted in Carr and Carr 1981, 17) concludes that nearly every European party, "known for not reading each other's journals and accounts," including Vlamingh in 1696, Grey in 1839, and McDouall Stuart in 1864, suffered cycad poisoning.

Nevertheless, some of the best-preserved records of Aboriginal uses of plants, such as zamia, come from the published accounts of explorers and settlers (for example, Grey 1841a, b, Moore 1846, 1884). Contrary to Beaton's assertion, these records indicate familiarity with the observations of previous explorers and naturalists, and document positive interactions between explorers and Aboriginal communities. Along the Arrowsmith River north of Perth, George Grey (1841a, 61), guided by Kaiber, was alerted to the toxicity of zamia: "Kaiber brought in some nuts of the Zamia tree; they were dry, and therefore in a fit state to eat." The rest of Grey's party indulged impetuously in insufficiently dried fruits, leading to "violent fits of vomiting accompanied by vertigo, and other distressing symptoms" (Grey 1841a). Grey's party exhibited what is now known as *zamia staggers* (Carr and Carr 1981).

Aboriginal cultures throughout Australia, including the Nyoongar of the South-West to whom zamia fruit is known as *by-yu*, have evolved strategies of detoxification, including roasting, soaking, and fermenting, or a combination of techniques, to convert the nut into a staple food resource. In the South-West, explorers and writers, including botanist James Drummond in 1839, naturalist and Quaker missionary James Backhouse in 1843, HMS Beagle officer John Lort Stokes in 1846, and chronicler J.E. Hammond in 1933, observed the significance of zamia palm to local Aboriginal people (Meagher 1974, 25). Lawyer George Fletcher Moore (1846) observed Nyoongar processing of the cycad fruit and its subsequent detoxification:

> This in its natural state is poisonous; but the natives, who are very fond of it, deprive it of its injurious qualities by soaking it in water for a few days, and then burying it in sand, where it is left until nearly dry, and is then fit to eat. They usually roast it, when it possesses a flavour not unlike a mealy chestnut; it is in full season in the month of May. It is almost the only thing at all approaching to a fruit which the country produces. (17)

Western Australian settler and creator of one of the first Nyoongar-to-English dictionaries, Moore (1846, 22) lists the term *djiriji* for the zamia as containing "a farinaceous matter, which, when prepared, has been used as sago, but is dangerous without preparation." *Gargoin* denotes the pit of the zamia fruit, "edible after being steeped in water or buried in moist earth for a time; but the kernel is considered unwholesome by some persons" (Moore 1846, 28). The complexity of the Nyoongar vocabulary surrounding zamia signifies its cultural importance as a foodstuff. According to Moore's dictionary, the Nyoongar differentiate between coastal species of zamia, such as *kundăgor*, and between the outer kernal *d-yundo* and the inner kernel *wi-dă* of the nut or *kwinin*.

The poisonous yet nutritious nuts symbolize the progressive understanding of the endemic flora of the South-West by settler society, a process of conciliation that continues with contemporary botanists who forward the revisioning of botanical conventions to provide models for how South-West species adapt uniquely to local ecological constraints (George 2002a, George 2002b, Hopper 1998). This brief account of the zamia palm illustrates the potential meeting of multiple narrative streams including poetic, scientific, and Aboriginal knowledges, each of which reflects complex sensuous histories. On the one hand, Nyoongar plant narratives refer to the edibility and palpability of flora with direct influence on human sustenance and cultural longevity. On the other, scientific observation is the apotheosis of visual denomination, leading to structured, hierarchical knowledge about the natural world and the relationships between species. What matters most in post-colonial Australia and elsewhere is the co-existence of narrative streams towards the possibility of dialogue between cultures and ways of knowing.

Nyoongar Conceptualizations of Plants

Before the relatively recent history of European colonization of Australia, there was (and continues to be) the cultural richness of 50,000 years of Nyoongar interaction with plants as food, medicine, tools, ornamentation, and totems. Nyoongar culture and the history of South-West flora are inseparably entwined. Plants have made spiritual and material sustenance possible while, conversely, Aboriginal people have ensured the longevity of plant populations (for example, see Hallam 1975). Joe Northover (1998, 40) expresses movingly the dialogical relationship between Nyoongar people and their land: "We don't have Cathedrals or built monuments to celebrate our culture, we have landscape and the very landscape is a reflection on us and we are a reflection on our landscape." In Nyoongar belief, plants belong to a spiritual landscape. However, this integrated spiritual and material alignment with local plants can often contradict the imperatives of managerialistic forestry practices or conservation science (McCabe 1998).

Through acts of sustenance, Nyoongar peoples have developed complex corporeal knowledge of flora that can deepen visual appreciation of plants as "beautiful flowers" or "sublime forests." Indeed, the histories of Aboriginal peoples throughout Australia are enmeshed with plant histories (Clarke 2007). Whereas scientific knowledge relies on universalized structuring that can exclude the embodied experience of the aesthetic (human) subject, Aboriginal plant narratives are predicated on edibility, palpability, aroma, and the intricate connections between the senses and eco-cultural meanings. Nyoongar history intertwines plants with the Aboriginal Dreaming, the complex stories and proscriptions that engender acts to ensure the sustained health and productivity of the land. Robert Bropho (1998, 31) states that "all the Dreaming stories are within the roots of that tree, coming from the ground and [the stories and the roots] can never be separated." Bropho (1998, 31) aligns human and tree bodies through a form of visceral empathy between species, but one ultimately founded on commiseration: "When I see those photos in the papers of the logs laying there with no limbs on them [I think] that's a body of a Blackman there from the neck down to his ankles and everything's been trimmed [...] that hurts me." A kindredness between plants and people points to the shared consequences for both, as organisms symbiotically occupying (whether we like it or not) ecological spaces. Ted Wilkes (1998, 45) observes that "the trees in the forest in the southwest of Australia have gone through exactly the same thing that Aboriginal people have gone through—annihilation, dispossession."

Traditional Nyoongar interactions with plants are exemplary of "embodied spatiality," a term proposed by environmental humanities scholars Deborah Bird Rose and Libby Robin (2004) to encapsulate the experience of physical connection through acts of sustenance, including the gathering of plant foods, medicines, and materials. The visual features of plants, however, can overlay (and sometimes override) deeper embodied cultural resonances. Dorothy Collard (1998, 34) reflects on the difference between Nyoongar traditional knowledge and the practices of modern forest management in relation to the belief in the restoration of cleared old growth forests: "[The forest] will never, never be the same. [It] might look good, with their eyes but the spirits [are] not there." While an ecosystem can be reconstructed visually, the soul or essence of the forest is irreplaceable. Furthermore, in terms of the consequences of clear-felling old growth forests, Mike Hill (1998, 18) alludes to the cultural interdependencies Nyoongar people have maintained with flora and the bodily histories that become endangered when ecosystems are altered and species likewise become endangered.

In the kwongan, edible roots, bulbs, and tubers have been culturally and spiritually significant, as elder Ken Colbung (1998, 53) observes: "If you ate the food that was around the area, and that was what you had to do is eat the food that was in the region where you went, your [...] magnetic being was more present." According to Colbung, interactions with wild foods resonate in this ways through bodily participation in the environment and the act of eating. Sense of place becomes palpable and embodied spatiality a phenomenon of taste. In fact, the edible wild yam (*Dioscorea hastifolia*), known as *adjikoh* or *ijjecka*, influenced the degree of sedentism of particular Aboriginal communities (Carr and Carr 1981, Hallam 1989, Hallam 1975). M.A. Bain (1975, 151) suggests that Nyoongar agricultural practices along the Irwin Valley south of Geraldton were focused on the cultivation of root crops: "The people in the vicinity of the Bowes River lived mainly on *ijjecka* root [...] it appeared [to settlers] to be a delectable and valuable yam, worth cultivating." Settler Lockier Burges was of the opinion that the variety of edible root crops, like the *ijjecka* growing prolifically and to great sizes, fostered a diet primarily of plant foods amongst the people of the Irwin Valley (Bain 1975, 46).

In the sandplains near the modern suburb of Wanneroo, Colbung (Graham 1990) demonstrates the edible and medicinal potential of the *bayu* (zamia fruits), *bera* (banksia flower), *boron* (bush red onion), *kojibut* (melaleuca balm), *kollookal* (pig face), *bayini* (wild fig), *mundar* (Christmas Tree), and *balga* (Xanthorrhea). Rather than being a sterile and austere place, the kwongan sandplains continue to nurture and sustain humanity. However, the nutritive and curative possibilities of species depend on bodily openness to the seasonal biorhythms linking people and plants (see Chapter 1). Specific procedures for detoxifying plant parts, such as the fruits of the *by-yu* cycad, have been generated as a dimension of cultural heritage (see Chapter 4). The bodies of plants dissolve into the bodies of people, sustaining enduring relationships of mutual benefit, as Sylvia Hallam outlined in her classic study *Fire and Hearth* (1975).

In reference to another root vegetable, George Fletcher Moore (1884/1978) recorded the usage of *konno* or *Platysace cirrosa*:

> I have discovered a bulbous root like a dark-coloured potatoe [*sic*], called by the natives *konno*, which I mean to endeavour to cultivate, and which may be very useful if it succeeds. The taste is something like the meat of a cocoanut [*sic*], or between that and a carrot taste. One specimen is as large as your fist. (301)

In 1842 in the Wongan Hills, the naturalist John Gilbert (quoted in Carr and Carr 1981) reported a harvest festival based on *konno*: "Their season of meeting in great numbers to dig the edible root called by them *Wargae* is now in full force." One-hundred and twenty-five years later, Sara Meagher (1974, 26) observed the collection of *karno* (*konno* in Moore's dictionary) near Mingenew: "The tubers are about half a metre below the ground and are dug up with a digging-stick [...] These tubers are available throughout the year and, besides being roasted in the ashes, are sometimes eaten raw to quench the thirst." These traditional narratives of flora include sensations of hunger, thirst, and sickness, and are therefore body-engaged and sensuous accounts.

Similarly, the endemic West Australia Christmas tree, known as *mundar*, *mudjar*, or *munji* in Nyoongar and *Nuytsia floribunda* to scientists, is a conspicuous South-West plant with very significant cultural meanings (Hopper 2010). The Christmas tree exhibits a variety of shapes and sizes and flowers around mid-December. Modern science classifies *Nuytsia* as an endemic mistletoe. As a root and rhizome hemi-parasite, it draws nutrients from a number of hosts but also possesses the ability to photosynthesize (Paczkowska and Chapman 2000). The parasitizing rootlets coming off the main roots of *Nuytsia* are so tenacious and well-designed that they have been known to burrow into underground utility lines.

In the 1930s, the ethnographer Daisy Bates (1992) recorded the connections between the spirit world of the Nyoongar and the Christmas tree:

> The tree-Moojarr, or Moodurt [...] was to the Bibbulmun the 'Kaanya Tree', 'the tree of the souls of the newly dead'. From time immemorial the soul of every Bibbulmun rested on the branches of this tree on leaving its mortal body for its heavenly home, Kurannup, the home of the Bibbulmun dead which lay beyond the western sea. (153)

Nuytsia facilitated the passage of souls to the after-world, and, as Bates claimed, the tree was consequently feared and avoided because of its power. However, other early ethnographers recorded the use of *Nuytsia* for food, water, and decoration, hence suggesting that there were variable cultural beliefs about the tree (Cunningham 2005, 223). Moore (1846, 80) described the Christmas tree as "Mut-yal, s.– Nuytsia floribunda; colonially, cabbage-tree. The only loranthus or parasite that grows by itself. Another anomaly in this land of contradictions. It bears a splendid orange flower." As the world's largest parasitic plant, the Christmas tree epitomized and still epitomizes the baffling growth habits of antipodean plant species. *Nuytsia* represented wholly the departure of Western Australian landscapes from European aesthetic norms (Chapter 2). Even the name *Christmas tree*, flowering in yellow as it does in the heat of November and December, runs contrary to the image of the evergreen Christmas tree brought indoors from the cold and deep snow of the English countryside.

Colonists and travelers reported a mix of admiration, fascination, fear, and disdain for *Nuytsia*. The tree in flower was first recorded by the crew of Dutch explorer Pieter Nuyts' vessel *Gulden Zeepaert* in 1627 (Cunningham 2005, 225). *Nuytsia* was assigned its scientific name in 1831 by Robert Brown, and the tree was referred to as "Fire Tree" amongst Swan River colonists (Lindley 1840, xxxix). In the journals of surveyors Alfred Hillman and Septimus Roe, *Nuytsia* indicated infertile country and was described disparagingly as part of the intolerable scrubbiness of the bush (Hopper 2010). In 1880, the peripatetic English botanical artist Marianne North travelled from Albany to Perth overland by carriage. North (1892) wrote in ecstatic terms about the hemi-parasitic tree: "I shall never forget one plain we came to, entirely surrounded by the nuytsia or mistletoe trees, in a full blaze of bloom. It looked like a bush-fire without smoke. The trees are, many of them, as big as average oaks in our hedgerows at home, and the stems are mere pith, not wood." North painted "Study of the West Australian Flame-tree or Fire-tree," now part of the botanical art collection at Kew Gardens in England. In form, the tree depicted in the painting appears closer to the European elm with its pleasant vase-like symmetry than most examples found in the Western Australian wild (to view North's painting, see http://www.kew.org/mng/gallery/764.html). Also in the late nineteenth century, Canadian novelist Gilbert Parker, travelling on the new railway, echoed North's appreciation of the pleasing composition of the bush during the spring months: "the yellow cabbage-tree flower [*Nuytsia*] is gleaming near, flanked by the white-and-green banksia, and a blossoming gum-tree is full of a regal beauty."

For some Nyoongar people, the Christmas tree has been considered a sacred plant connected to the souls of the deceased. But, importantly, the tree has also been a food. The records of settlers and ethnographers point to a combination of spiritual beliefs and material practices surrounding *Nuytsia*. Writing in the 1880s, Ethel Hassell (1975, 26) recorded the use of *Nuytsia* root as a candy: "They [the Nyoongar people] gave me one of the roots to taste, telling me it was called *mungah*. The outer skin was pale yellow but easily stripped off leaving a most brittle centre tasting very like sugar candy." A ghoulish creature called a *gnolum*, in the form of a very tall, very thin man, enticed boys away by offering them the roots of *mungah* (Hassell 1975, 65). Bates (1992, 153) recorded the view of *Nuytsia* as a home for disembodied spirits; the Nyoongar "did not fear the tree;

they loved it, but held it sacred for its spiritual memories. The souls of all their forbears had rested on the spirit tree on their way to Kurannup." In contemporary times, Nyoongar Elder Noel Nannup explains in an interview:

> *Moodjar*, the Christmas Tree, is one of the only flowers late in the [calendar] year when the easterly winds are blowing. The belief was that a spirit of the deceased person sat on the tree until it flowered. Then the spirit moved on to the spirit world, and that would be in conjunction with the easterly winds and fire, so natural fires burned throughout that time of year. The easterly wind would take the spirit out over the sea. (personal communication, N. Nannup, July 21, 2010)

Noel offers a more complex view of the sacredness of *moodjar* as interconnected to easterly winds and the fire season. According to the Nyoongar calendar, the onset of *birok*, the season encompassing December and January, corresponds to the flowering of *mudja* along with a suite of ecological indicators that prompted the movement of people to coastal regions (Chapter 1).

Conclusion: Botanical Narratives and the Senses

Springing from Thoreau's sensuous explorations of North American flora, this chapter has presented *multiple narrative streams* as a way to conceptualize the possible convergence of the diverse narratives of plants and their aesthetic appreciation. Science has a functional view of plant life aligned to its classificatory prerogatives and taxonomic aesthetics. In contrast and complement, traditional Nyoongar stories of plants revolve around corporeality through the seasons toward the production of ongoing cultural and spiritual meanings. Narrative hybridity returns to the sensuous and intellectual histories of interactions between flora and culture—living histories combining science with other ways of knowing. Through narrative streams, the writing of nature and natural history becomes oriented toward transdisciplinary forms of knowledge spanning science, poetic thought, Aboriginal traditions, and personal experience. That such streams of knowledge meander together in writing of all forms toward creative possibilities is more important than their epistemological differences or aesthetic incompatibilities. As suggested by Judith Wright at the beginning of this chapter, poetry has the potential to bring metaphor and myth to our understandings *of* and relationships *with* plants. This assertion is especially germane to the botanical poetry of the South-West of Western Australia, the subject of Chapter 6 of *Being With*.

Chapter 4: Cultures of Flora: Conserving Perth's Botanical Heritage through a Digital Repository

Introduction

How should we record and preserve botanical narratives and their associated heritage? How does heritage relate to the concept of being with the botanical world? With the increasing loss of bushland to urban development worldwide, the conservation of plant diversity is a challenge faced by cities across the globe. What is obscured in urban botanical conservation practices is that, in protecting biodiversity through a focus on endangered species and their habitats, cities conserve—concurrently—the heritage values of their plants. FloraCultures develops a holistic framework for plant-based cultural heritage safeguarding that incorporates the multiple heritage values of Perth, Western Australia, flora. Drawing from current theories in ethnobotany, heritage conservation, and digital design, this chapter outlines a model for conserving the heritage of Perth's flora. FloraCultures presents a multimedia repository to consolidate and make publicly accessible the cultural-botanical values of Kings Park and Botanic Gardens, the premier site for botanical conservation and education in the State. The project collates a range of information, dispersed across a number of cultural traditions and materials, and including Aboriginal, colonial European, and recent immigrant knowledges.

Background to the Project

FloraCultures is a 2013–15 pilot project in development with Kings Park and Botanic Garden in Perth, WA, and funded by Edith Cowan University's Early Career Researcher grant scheme. The project aims to develop a model for documenting the plant-based cultural heritage of 30–50 indigenous species occurring in the Kings Park bushland (Figure 4). The FloraCultures initiative (www.FloraCultures.org.au) integrates archival and digital design techniques, creating a unique web portal of potential interest to a range of users—from first-time tourists and amateur naturalists to heritage consultants and environmental conservationists (Figure 4). The initiative reflects the belief that research into environmental heritage (defined broadly to encompass natural and cultural heritage and tangible and intangible theory) is integral to the conservation of flora and fauna in their ecological habitats. The project stresses that the appreciation of biodiversity for its cultural significance helps to sustain broader conservation values.

Figure 3. This aerial view of Kings Park and surrounding suburbs shows the extent of the bushland area in brown.
Source: Kings Park and Botanic Garden
(http://www.bgpa.wa.gov.au/kings-park/maps/aerial-view)

FloraCultures responds to the increasing pressures forced upon the greater urban environment in which Kings Park is located. The project develops knowledge and practice of plant-based cultural heritage conservation in Perth, one of the world's most ecologically unique cities. According to botanist and outgoing director of the Kew Royal Botanic Gardens, Professor Stephen Hopper, Perth is one of the world's most biodiverse urban areas, particularly with respect to plant life (quoted in Perth Biodiversity Project n.d., 1). However, the continued loss of remnant bushland to development, disease, and disregard poses a threat to botanical conservation, while also challenging the long-term environmental and cultural sustainability of Perth and other cities globally (Farr 2012, Hostetler 2012, Knapp 2010). In considering the conjunction between natural heritage and biodiversity conservation, the challenge becomes more pressing. The continuity of plant-based cultural heritage (the acronym PBCH will be used intermittently in this chapter) depends intrinsically on the survival of floral and faunal species (Pardo-de-Santayana, Pieroni, and Puri 2010, 1). Moreover, for the purposes of this discussion, synonymous terms for plant-based cultural heritage will include "botanical heritage," "biocultural heritage," and "ethnobotanical heritage" (Pardo-de-Santayana, Pieroni, and Puri 2010, 1, 5).

Figure 4. The FloraCultures holding page features Marianne North's painting "Study of the West Australian Flame-tree or Fire-tree" (circa 1880). Royal Kew Gardens generously granted permission to reproduce North's image on the holding page.
Source: Adrianne Barba, Horse & Cart Design.

Redefining, Documenting, and Conserving Urban Plant-based Cultural Heritage

The conservation of plant-based cultural heritage in Perth and other Australian cities is endangered by a variety of environmental and social factors, including urban bushland development (McKinney 2005), the loss of plant species to disease (Shearer et al. 2007), and demographic and social shifts in urban areas (Girardet 2008). In different regions of Australia, the conservation of plant-based cultural heritage has been initiated through its documentation in print-based works. Key Australia-wide publications, such as *Aboriginal People and their Plants* (2007), unfortunately, include only abbreviated references to Western Australian species and tend to exclude non-Aboriginal plant-based heritage information (Clarke 2007, 80-81).

Research into plants, people, and heritage production is normally situated within ethnobotany, an interdiscipline combining anthropology and botany and, historically, prioritizing Aboriginal Australian knowledges of plants (Clarke 2007, Hsu 2010, Isaacs 1989). While comprehensive ethnobotanical surveys exist for certain regions of Australia (for instance, see Latz 1995 for Central Australian documentation), relatively little published research on plant-based cultural heritage is available for Western Australia where information remains unrecorded, scarce, and dispersed across a range of sources (key works include Daw, Walley, and Keighery 1997, Meagher 1974). In other words, the problem of plant-based cultural heritage conservation in Western Australia is compounded by the dispersed nature of knowledge, occurring over a gamut of textual, visual, and oral sources and ranging across cultural traditions, including Aboriginal and non-Aboriginal.

Moreover, definitions of heritage tend to prioritize material or tangible heritage, comprising physical substances (of living plants), cultural artifacts (made from plants), and natural sites (containing plants) (Denes 2012, 165, Sorensen and Carman 2009). As a regional example, the Heritage of Western Australia Act 1990 emphasizes structures and places of consequence, defining cultural heritage as "the relative value which that place has in terms of its aesthetic, historic, scientific, or social significance, for the present community and future generations" (Western Australia 2012, 3). Using this definition as its basis, the term *botanical heritage* would imply the one-sided conservation of tangible plant-based cultural heritage—including significant living plants (e.g. banksias), their broader habitats (e.g. banksia woodlands), and the enduring materials produced from them (e.g. colonial-era furniture made from banksia wood).

FloraCultures aims for a more inclusive conceptualization and practice of natural heritage in relation to plants—one that equally considers, consolidates, and conserves diverse forms of heritage. The first form addressed by the heritage initiative is knowledge of Western Australian plants as food, ornamentation, medicine, and fiber (Clarke 2007, Cotton 1996, Hoffman and Gallaher 2007, Martin 2004). The second form is knowledge of plants as literary, artistic, and historical objects (Mahood 2008, Ryan 2012a, Seddon 1997, 2005). The third form is knowledge of plants as sources of community memory, cultural identity, and personal well-being, largely derived through ethnographic and oral history techniques (Hitchings 2003, Hitchings and Jones 2004, Ryan 2012a, Trigger and Mulcock 2005).

Bridging Tangible and Intangible Plant-based Cultural Heritage

FloraCultures involves a holistic framework for plant-based cultural heritage safeguarding that incorporates the conservation of the tangible and intangible heritage values of plants (Leyew 2011, 157) with specific focus on Perth's flora. The research expands Western Australian environmental heritage theory and practice through the application of a syncretic framework to plant-based cultural heritage. Intangible heritage (ICH or "intangible cultural heritage") refers to "forms of cultural heritage that lack physical manifestation. It also evokes that which is untouchable, such as knowledge, memories and feelings" (Stefano, Davis, and Corsane 2012, 1, Isar and Anheier 2011). The 2003 UNESCO *Convention for the Safeguarding of the Intangible Cultural Heritage* lists five manifestations of intangible heritage: "(1) oral traditions and expressions, including language as a vehicle of [...] intangible cultural heritage; (2)

performing arts; (3) social practices, rituals and festive events; (4) knowledge and practices concerning nature and the universe; and (5) traditional craftsmenship" (UNESCO 2003, 2). Many plant-based cultural heritage researchers regard the UNESCO convention as a "turning point [for] ethnobiology and ethnosciences [in] recognising all orally transmitted traditional knowledge systems" and the need for further research into their safeguarding and conservation (Pardo-de-Santayana, et al. 2010, 11).

The category "knowledge and practices concerning nature" is of particular relevance to FloraCultures. A plant-based heritage resource, linking intangible and tangible heritage, has not yet been envisioned, designed, or produced for Western Australia. In comparison, and particularly since the 1970s, scientific knowledge of the State's plants has been increasingly promoted, funded, and consolidated. Technical knowledge of Perth-area plants is available through online sources, such as FloraBase, and key print publications, most notably *Flora of the Perth Region* (Marchant et al. 1987) and *Perth Plants* (Barrett and Pin Tay 2005). Developed by the Western Australian Herbarium and released online in 1998, FloraBase is the "authoritative source" for plant knowledge in the state (Department of Environment and Conservation 2012). FloraBase provides a free, open-access tool for researching topics such as the scientific and common names of plants; the conservation status of different species; the taxonomic features aiding identification; and species distributions. Moreover, the database includes color photos of the major features of most of the 10,000 Western Australian plants. In its breadth and completeness, FloraBase has contributed substantially to the promotion of plant-based scientific heritage in WA over the last fifteen years. The FloraCultures portal, once developed into an online tool, will complement FloraBase, fostering a productive dialogue between the botanical arts and sciences in the State.

While a comprehensive resource for Western Australian plant-based cultural heritage has not been published or developed, noteworthy examples of digital platforms for PBCH promotion do exist elsewhere. The Native American Ethnobotany Database, managed by the University of Michigan (USA), highlights the uses of plants as food, fibers, and medicines in Aboriginal North American societies (University of Michigan 2003). For example, a search for "maple" yields multiple ethnobotanical references, found in historical documentation, to the plant's use as medicine, material, and food by Native American people. Moreover, the website Plant Cultures, supported by Royal Botanic Gardens, Kew, showcases the uses of 25 prominent South Asian plants, including banyan, cardamom, garlic, ginger, mango, and turmeric. The website is eclectic in its knowledge base and inclusive in its selected media, bringing together "historic images from museums and libraries, well researched information, contributions from members of the public, and carefully chosen links to other web resources" (Kew 2005). For Western Australian plants, limited ethnobotanical information—such as interview material and Aboriginal nomenclature—is included in the website Nidja Beeliar Boodjar Noonookury Nyininy or A Nyungar Interpretive History of the Use of Boodjar in the Vicinity of Murdoch University, Perth, Western Australia (Murdoch University 2003).

FloraCultures Methodology: Combining Archival and Digital Techniques

The FloraCultures initiative draws from these online examples in order to fill a Western Australian gap with a searchable archive and educational tool. The project will result in web-based access to the cultural heritage of Perth's indigenous flora, defined broadly as trees, shrubs, and herbaceous plants existing in the Swan River area during early nineteenth-century European settlement. Once completed in 2014, the pilot website will offer a user-friendly platform for the documentation and promotion of urban Australian plant-based cultural heritage. In developing a comprehensive model, the FloraCultures pilot collates Aboriginal, colonial European, recent immigrant, local conservationist, and seasonal visitor knowledges of significant Kings Park species. Rather than developing a print-based ethnobotanical resource—comparable, for example, to *Bushfires and Bushtucker: Aboriginal Plant Use in Central Australia* (Latz 1995)—for Perth and Kings Park, FloraCultures will build an easily updatable digital archive accessible to a wide audience and incorporating non-Indigenous and Indigenous materials.

The 2013–15 pilot puts forward a dual archival and digital approach to heritage conservation with respect to the highly visible urban botanical reserve of Kings Park and Botanic Garden (Barrett and Pin Tay 2005, Erickson 2009). The site is significant for Perth biodiversity because, on average, 6 million people visit Kings Park per annum (pers. comm., M. Broderick, August 14, 2012). In addition to its visibility, centrality, and accessibility, many of Western Australia's iconic and endangered species are represented in the Kings Park bushland reserve (Figure 3). The digital repository maximizes public accessibility to the multiple heritage values of 30–50 iconic Kings Park plants, currently in the process of being selected in consultation with Kings Park personnel. FloraCultures is significant for cultural heritage, botanical conservation, and environmental education in Western Australia, generally, and Perth, specifically. Moreover, the project expands the knowledge bases and methodologies of ethnobotany, environmental education, sustainability studies, heritage studies, and digital heritage conservation in reference to this region.

Incorporating tangible and intangible heritage frameworks, the concept of plant-based cultural heritage provides the analytical underpinning of the FloraCultures initiative. Drawing from these conceptual positions, the project engages two interlinked approaches: (a) the identification of the tangible and intangible plant-based cultural heritage of Kings Park as expressed in textual, visual, and oral sources; and (b) the archival and display of the research findings for public consumption through a freely accessible and multimedia (i.e. text, images, audio, and video) web resource. Hence, the methodology consists of Phase I and Phase II during which distinct but related activities address three research questions: (a) Which Kings Park plants have the strongest heritage values; (b) What kinds of heritage information are most relevant to conserving the overall PBCH of Kings Park; and (c) How can a holistic model of PBCH (including Indigenous and non-Indigenous knowledges as well as tangible and intangible heritages) address gaps in environmental heritage conservation and promotion in the State?

Phase I uses an archival methodology beginning with the documentation of plant-based cultural heritage references in WA-based sources. At the same time, intangible heritage is being recorded through interviews with individuals with "knowledge, memories and feelings" (Stefano, et al. 2012, 1) of Kings Park plants. Whereas Phase I assembles a portrait of plant-based cultural heritage in textual, visual, and spoken sources, Phase II focuses on the production of a digital archive, accessible to a variety of users (Figure 4). The archive development showcases current digital design and user interface theories and practices (for example, Cross 2011, Meinel 2011), including the use of focus groups, in order to maximize the impact, applicability, and visibility of the project and its outcomes.

In greater detail, FloraCultures Phase I, "Documenting Plant-Based Cultural Heritage," focuses on researching the representations of iconic Kings Park plants in historical texts (explorers' diaries, newspaper articles, and the published accounts of settlers, visitors, and naturalists) (for example, Millett 1872, Moore 1884/1978, 2006); literature, including novels (e.g. Lawrence and Skinner 1924), poetry (e.g. Alexander 1979, Kinsella 2005), and short stories; visual art (sketches, paintings, illustrations, and photography) (e.g. Nikulinsky and Hopper 1999, 2005); music (e.g. Grainger 1985); and film (e.g. Graham 1990). The references provided here are representative of the spectrum of materials that is being curated for inclusion in the repository. Between 2013 and 2015, archival research is being conducted at the J.S. Battye Library of WA History; State Records Office (SRO); Private Archives Collection at the State Library; State Library's Pictorial Collection; Trove Digital Archive of the National Library of Australia; WA Museum; WA Historical Society; Edith Cowan University Library; UWA Library; and relevant private archives and collections at Kings Park and Perth suburbs. The selection of archival material focuses on prominent plants—such as kangaroo paws discussed in the next section—that figure recurringly in archival and cultural sources, and are integral to the plant-based cultural heritage of Western Australia. Additionally, the curatorial process attempts to balance textual, visual, and audio materials of strong heritage value as per the constraints of the online format and copyright restrictions.

Moreover, throughout Phase I, Aboriginal Australian PBCH is being identified in various textual, visual, and oral accounts (e.g. Bates 1992), assessed by an Indigenous Heritage Consultant and, subsequently, included or excluded. Additionally, the Phase I interview component documents the spoken accounts of Kings Park conservationists and other individuals with long-standing links to the area's flora. I record this intangible heritage information with a hand-held digital device and subsequently transcribe the conversations. The interviews are approximately one-hour in duration, semi-structured in format, and focused on recording the "knowledge, memories and feelings" (Stefano, et al. 2012, 1) of 6–10 interviewees, identified during the course of the research (for a fuller treatment of the concept of "botanical memory," see Ryan 2012, Chapter 8).

FloraCultures Phase II, "Disseminating Plant-Based Cultural Heritage," focuses on digital design and development. This phase begins with focus groups in 2013. In keeping with user-engaged design principles, four hour-long meetings with Kings Park volunteers, guides, and associates help to establish: (a) the digital archive features of most potential benefit to users; (b) species to feature within the archive; and (c) interviewees to contact. The focus groups have produced a set of recommendations (including a list of plant species and potential interviewees) that will be incorporated into the design process. The construction of the web portal will mostly take place between January and June 2014. I work closely with a professional Digital Development Consultant to ensure satisfactory incorporation of the focus group recommendations. After the production of a prototype and between October and December 2014, I will then organize training sessions for Kings Park personnel and the general public. The sessions will address: (a) the concept of plant-based cultural heritage; (b) the use of the information for educational purposes; and (c) the navigation of the web resource. The portal's promotion, maintenance, and possible migration will be addressed in detail at the end of 2014 and early 2015.

Gift of Blood: Catspaws and Kangaroo Paws (*Anigozanthos* spp.)

In order to demonstrate the kinds of "data" that will be featured in the FloraCultures web archive, I will briefly sketch in this section the plant-based cultural heritage of the catspaws (*Anigozanthos humilis*) and kangaroo paws (*Anigozanthos manglesii*) using the combined tangible and intangible heritage framework. In doing so, I will touch upon the three categories of plant-based cultural heritage that distinguish FloraCultures from other ethnobotanical resources, both print or online: (a) knowledge of plants as food, ornamentation, medicine, and fiber; (b) knowledge of plants as literary, artistic, and historical objects; and (c) knowledge of plants as sources of community memory, cultural identity, and personal well-being. The research process reveals, for example, that some plant species have more documented uses as food than citations as literary objects; hence the archival information is not symmetrical across species, but rather weighted in most instances toward one or two heritage categories.

Both catspaws and kangaroo paws belong to the Haemodoraceae—the botanical family name derived poetically from *haima* for "blood" and *doron* for "gift." The catspaw genus stems from the descriptive terms *anisos* for "unequal" and *anthos* for "flower," alluding to unsymmetrical lobes of the perianth; and *humilis* for "low-growing." Catspaws are also known by a number of common names including small catspaws, dwarf catspaws, Mogumber catspaws (for the sub-species *chrysanthus*), and tall cat's paw (for the sub-species *grandis*). In terms of traditional knowledge of human uses of catspaws, the plants have been prepared by Nyoongar people—the Aboriginal people of the South-West, including Perth (see Chapter 1)—as a dye. The compound haemocorin causes the reddish-orange color of the roots—an aesthetic property that is typical of members of the Haemodoraceae family. Moreover, haemocorin is medically significant and therefore a biochemical component of an embodied aesthetic of these plants; the substance is currently being researched for antitumor and antibacterial properties. Regarding the second category of plant-based cultural heritage, knowledge of plants as literary, artistic, or historical objects, in 1840, the enterprising British botanist John Lindley commented glowingly that "there is a dwarf species still handsomer than they [*A. manglesii* and *flavida*], in consequence of the compactness of the flowers and the short neat foliage; this *A. humilis* would be a handsome addition to our gardens' (Lindley 1840, xlvi). Due to its "handsome" appearance and attractive coloration, the catspaw has been depicted in seminal works of Western Australian botany and in the works of notable artists such as Australian painter Marian Ellis Rowan (1848–1922), both of whom will be represented in FloraCultures pending copyright permissions.

As suggested above, FloraCultures begins with the exploration of plant-based cultural heritage with the common, scientific, and Nyoongar names for plants. The scientific name for kangaroo paws (*Anigozanthos manglesii*) also derives descriptively from *anisos* for "unequal" and *anthos* for "flower," again for the unequal perianth lobes. However, the kangaroo paw species name honors Captain James Mangles (1786–1867), botanical enthusiast, seed trader, and cousin of Lady Ellen Stirling, wife of colonial governor James Stirling. Mangles visited the Swan River Colony in 1831 where he also met and began correspondence with Georgiana Molloy (1805–1843), Western Australia's first female botanist and one of the first botanical collectors and seed traders in the colony. The common names for kangaroo paws include red and green kangaroo paw, Mangles kangaroo paw and common green kangaroo paw. Multiple Nyoongar names for kangaroo paws are documented: *kuttych, kurulbrang, krulbrang, nollamara* (coincidentally also the name for a Perth suburb), and *yonga marra*. For Nyoongar people, kangaroo paws are known traditionally as food sources; the tender, starchy, and nutritious rhizomes were consumed before the emergence of the flower. Kangaroo paws also figure prominently into colonial Australian history. In 1834, botanist David Don published in *The British Flower Garden* the first formal description of a cultivated Mangles kangaroo paw: "This singularly beautiful species of *Anigozanthos* was raised in the garden at Whitmore Lodge, Berks, the seat of Robert Mangles, Esq. from seeds brought from Swan River by Sir James Stirling, the enterprising governor of that colony, by whom they had been presented to Mr. Mangles" (Don 1835, 265).

Figure 5. This kangaroo paw sculpture at Kings Park and Botanic Garden demonstrates the importance of the species to Western Australian identity.
Source: Author.

Images of kangaroo paws help to define and communicate contemporary Western Australian character and culture (Figure 5). In the 1860s, the English watercolorist G.C. Fenton included a painting of a kangaroo paw in his *Sketchbook* (http://nla.gov.au/nla.pic-an5836980). In 1960, the kangaroo paw was selected as the WA state floral emblem by Premier David Brand. An image of a kangaroo paw also frames the crown in the Western Australian Coat of Arms. The blazon reads: "And for Crest: On a Wreath Or and Sable The Royal Crown between two Kangaroo Paw (*Anigosanthos* [sic] *Manglesii*) flowers slipped proper." Kangaroo paws are nearly synonymous with wildflower tourism and have been used on impressively designed promotional posters since the 1950s (see, for instance, http://trove.nla.gov.au/version/7079089). The species also figure into the state's literary history (for example, see William Hart-Smith's poem "Kangaroo-Paw") (http://andrewlansdown.com/fellow-writers/william-hart-smith/).

Conclusion: The Practice of Botanical Heritage

FloraCultures contributes to the conceptualization, consolidation, and conservation of WA plant-based cultural heritage. Through the digital platform, the richness and importance of Perth's flora will be showcased to national and international audiences. The strength of FloraCultures is that it works across heritage streams and across different cultural traditions and historical periods. FloraCultures provides a plant-based cultural heritage model for Australian cities through a combination of innovative methodology (archival and digital approaches) and accessible technology (the web archive itself and future social media integrations). Further along in its development, the archive will solicit public contributions of PBCH materials (e.g. photos, stories, diary excerpts, etc.). I believe that the outcomes of the project will be valuable to diverse parties, including educators; environmental researchers; historians; cultural heritage managers; biological scientists; community conservationists; wildflower tourists; proprietors of botanical tourism; the mining sector and industries with an ecological restoration component; and others identified in focus groups. Multiple economic, environmental, and cultural benefits will accrue through the promotion of Western Australian ecotourism assets and the development of biocultural heritage management models; urban biodiversity conservation; and sustainability theory and practice in Australian cities.

Once the prototype has been completed, FloraCultures will set a precedent for promoting PBCH. As such, the project has potential benefits for plant-based cultural heritage policy and practice in Western Australia. In forwarding new notions of PBCH, FloraCultures is poised to impact heritage conservation over time. Firstly, the research expands ethnobotanical practice in Australia by incorporating non-Indigenous knowledges and intangible cultural heritage theories. Secondly, the digital platform and associated methodologies will set a precedent for further research into PBCH, presently under-developed and, arguably, threatened in Western Australia. Thirdly, the ethical and copyright issues—particularly regarding Nyoongar intangible heritage and archival material respectively—navigated during the project, especially in 2014, will be of relevance to subsequent Australian cultural heritage research. Importantly, the prototype will be poised for social media integrations. For example, the development of an iPhone application would connect the repository to a broader user base while enhancing its educational possibilities.

FloraCultures will foster further academic engagement with Kings Park, to promote the heritage value of its plants and to contribute to a broader recognition of Western Australian plant life—two of the institution's core aims. FloraCultures also contributes uniquely to two Australian National Research Priorities: *An Environmentally Sustainable Australia (*Goal 5: Sustainable Use of Australia's Biodiversity) and *Promoting and Maintaining Good Health* (Goal 4: Strengthening Australia's Social and Economic Fabric). The Project develops strategies for conserving the heritage values of Australia's terrestrial biodiversity. By foregrounding the long-term social and cultural significance of biodiversity to Australian natural heritage (both tangible and intangible), the outcomes will contribute to the conservation of biodiversity for both its *inherent* value and *economic* benefits, especially to the tourism sector. Moreover, the project aligns with Australia's National Landscapes Program and the Tourism 2020 initiative for developing promotional strategies for natural and cultural heritage. As the fastest growing segment of Australian tourism, ecotourism (including wildflower tourism) will be furthered. FloraCultures contributes to the second NRP by strengthening Australia's social and economic fabric through sustained community engagement. In building the capacity of communities for heritage conservation, the project will enhance the potential for *healthy*, *productive*, and *fulfilling* lives in Western Australia. Alongside these potential practical benefits, the aim of FloraCultures originates in a poetics of plant life and a philosophy of being with.

Chapter 5: Reading Botanical Aesthetics: Embodied Perceptions of Perth's Flora, 1829 to 1929

Introduction

The interconnections between plants, aesthetics, and sustainability are crucial for environmental planners and philosophers in Perth and elsewhere to consider. The problem is that aesthetics, as theorized in Western discourse, narrows the human experience of flora to the perception of visual beauty. This has potentially negative consequences for current and future human-plant interactions in urban settings and for the long-term conservation of bushland flora, as discussed in Chapter 4. The ideas of Kant, Schachtel, and Porteous are based on hierarchies that place the "lower" senses of taste, smell, touch, and proprioception in an inferior position to sight and hearing. An *aesthesis* of plants—or what I theorize as *floraesthesis*—opens human experience to the sensations of flora, to the intimacies of contact with nature that, hopefully, promote sustainability, conservation, and the ethical consideration of the botanical world. Floraesthesis—a sensory being with plants—is the conceptual framework I use to analyze a segment of written material collected as part of a long-term heritage archiving project FloraCultures (Chapter 4). Reading historical documents through the lens of aesthesis engenders a multi-sensory perspective on colonial Western Australian attitudes toward indigenous flora during the period following settlement, roughly circumscribed for this discussion as 1829 to 1929. Not all of these attitudes were linked to the clearance of Perth's trees, shrubs, and herbs. On the contrary, the broader physical distribution of plants in the colonial environment and the severe lack in the availability of essential goods necessitated embodied interactions with flora as food, medicine, fiber, and ornamentation. Indeed, threaded throughout these written accounts are early messages of conservation which we can learn from today.

Sustainability and Being with Plants

> There's a shrubby plant in blossom just now that lends a great deal of beauty and variety to our bush undergrowth. It is especially beautiful when long shafts of morning sunshine filter through trees and bushes, diversifying the monotony of flower-gemmed green with charming light and shade effects of golden sunlight and purple shadow patches.
>
> —In reference to Blueboy (*Stirlingia latifolia*) (The West Australian 1924, September 19, 6)

Sustainability—ever-more a contested term (Thompson 2010, 196-214)—can be defined as the meaningful and dynamic long-term equilibrium between environmental and social, human and nonhuman, sentient and non-sentient "things" (see Chapter 1) co-existing in a physical space. Plant life and human relationships to the botanical world are crucial dimensions of sustainable communities and ethical ecocultural practices. However, the role of plant life in the theory and practice of sustainability is problematically limited to the utilitarian discourses of sustainable agriculture (Tuteja 2012), food security (Wright 2012), organic farming (Burnett 2008), urban gardening (Reid 2012), sustainable forestry management (Kitayama 2012), and ideological debates over invasive plants and their impacts on ecosystems and indigenous species (Coates 2006). Such discourses exemplify an overwhelmingly limited anthropocentric perspective on the botanical world that largely disregards its other values—most importantly a plant's intrinsic right-to-exist (Hall 2009), and the metaphysics of the plant world (Marder 2013).

Despite the utilitarian emphasis, plants figure in multiple ways to the community-based, social, emotional, and spiritual facets of sustainability, for example, by supporting diverse dimensions of human wellbeing (Cotton 1996), contributing to the formation of cultural, regional, or place-based identity (Trigger and Mulcock 2005), and defining the aesthetic characteristics of a region, city, town, place, or site (Ryan 2012a). While current research into sustainability acknowledges the functional role of plants, few studies foreground the interrelationships between aesthetics, sustainability, and flora. Even fewer foreground the ways in which respectful interactions with plants, founded on human aesthetic values, can sustain social systems. The biogeographical context of the South-West Botanical Province—a biodiversity "hotspot" of international renown, with more than 8,500 species of plants and a 35% rate of floristic endemism (Hopper 2004), as outlined in the Preface—offers an exceptional case study for exploring the connections between aesthetics, sustainability, and flora. By virtue of its physical location in the larger South-West region, Perth can be considered one of the world's most ecologically diverse cities. The late writer and ecologist George Seddon advocated the cultural, social, and historic significance of WA plants in the seminal publications *Sense of Place* (1972) and *The Old Country* (2005). However, cultural inflections of the South-West's botanical diversity appear sporadically in historical writings during the first one-hundred years following the founding of the Swan River Colony (1829).

Inspired both by Seddon's writings and Perth's unique yet fragile flora, I began in 2013 initial work on an ambitious long-term project titled "FloraCultures," described in detail in Chapter 4, with seed funding from Edith Cowan University (www.FloraCultures.org.au). To expand on the discussion in the previous chapter, FloraCultures combines "design thinking" (Plattner 2012) and digital heritage conservation techniques (MacDonald 2012) in developing an online, open-access repository of WA plant-based heritage content. The project is working towards a broad-based, user-friendly, multi-cultural, and multimedia framework for botanical heritage conservation. In collaboration with Botanical Gardens and Parks Authority (BPGA), FloraCultures entails a small pilot study of plants found naturally in the bushland of Kings Park in Perth. Once the pilot project is released in 2014, the online repository will promote the overlays between cultural heritage and indigenous plants through a spectrum of content: Aboriginal Australian uses and beliefs, explorer journals, settler diaries, visitor accounts, poetry, literary works, paintings, photography, music, and oral histories. This chapter analyzes one component of the heritage archive: a cross-section of written material retrieved from the Battye Library, the digitized newspaper archives of the National Library of Australia (NLA), and other WA-based historical collections. These writings (mostly from newspaper and journal articles, in this instance) point to a sophisticated awareness of Perth's bushland flora among the city's burgeoning human population. In order to contextualize these excerpts, references will be made to published, non-fiction accounts of settlers (Moore 1884/1978) and visitors (Armstrong 1979) to the South-West region, 1829 through 1929.

An Embodied Aesthetics of Plants: Sensation and the Allure of the "Lower" Senses

Aesthetics has been historically associated with the visual appreciation of artworks (for example, see Berleant 2005). However, the aesthetics of art and nature share a common preoccupation with the appearance of objects, with vision as the principal sense, with distance as the habitus of interaction, and with analytical reflection as the mode of judgment (Ryan 2012a, Chapter 4). Accordingly, in order for a plant (tree, shrub, grass, wildflower, or herb) to become a proper object of aesthetics (i.e. to be judged in aesthetic terms), the visible qualities of form (symmetry, harmony, gracefulness, vastness, magnitude) and color (brightness, tone, contrast, homogeneity) must be manifest in the plant. The species otherwise falls partly or wholly outside of the conventions of aesthetics as conceptualized in Western thought. In *Aesthetic Theory*, Theodor Adorno (1984, 91) propounds that aesthetics, particularly since the work of the German philosopher Friedrich Wilhelm Schelling (1775–1854), has been concerned with art rather than nature. Adorno, reflecting a Kantian orientation, maintains that the aesthetics of art and the aesthetics of nature are interrelated; both are concerned with appearances, images and the assessment of beauty. Nature, in terms of aesthetics, is "perceived as [the] appearance of the beautiful and not as an object to be acted upon" (Adorno 1984, 97). As with an object of art, beauty in nature instigates analytical reflection, scientific rationality, and objective distance. Echoing Kantian disinterestedness, Adorno asserts that "the beautiful in nature is that which appears greater when seen from a distance, both temporally and spatially" (Adorno 1984, 104). Nature can be an object of aesthetics, but only if conflated with art, that is, only on *art's terms*.

Immanuel Kant's cognitive model of the senses has played a fundamental part in the Western notion of aesthetics and our regard for plants as beautiful and aesthetically worthy of moral consideration and, in some instances, worthy of habitat or heritage conservation. Kant's model centers on "empirical perception," otherwise known as object cognition—the process of objectification in which information about objects in the world (i.e. those things external to our bodies) is acquired as cognitive data. The senses that can deliver the most information about objects, subsequently, form the basis for a taxonomy, ranked from the lower (smell and taste) to the higher senses (touch, hearing, and vision). In *Anthropology from a Pragmatic Point of View*, Kant splits the "vital" senses from the "organic" senses. For Kant, changes in temperature, pressure, or emotional states engage the vital senses and "penetrate the body to the center of life" (Kant 1978, 41). For example, proprioception, as a vital sense, is the perception of one's muscles and joints as one moves through space. The five organic senses are either objective or subjective. For Kant, the objective senses are touch (*tactus*), sight (*visus*) and hearing (*auditus*); the subjective senses, taste (*gustus*) and smell (*olfactus*). The objective senses are empirical, leading to knowledge of an object (e.g. its shape and size); the subjective senses register impressions directly on or near the mediating organ (e.g. the tongue, stomach, nose, and lungs) and are associated with pleasure rather than cognition. Subjective sense involves variable responses between individuals; objective sense underpins the science of the senses because it results in perceptual consistency between individuals (i.e. form and color can be measured empirically and communicated). Kant explicitly links sensation to objectification: "all together they [the five senses] are senses of organic sensation which correspond in number to the inlets [nose, tongue, ears, eyes and skin] from the outside, provided by nature so that the creature is able to distinguish between objects" (Kant 1978, 41).

The Kantian model of sensation requires the cleavage between the objective (higher) and subjective (lower) senses. The difference between the higher and lower senses corresponds to the difference between surface perception and inner sensation. As such, sensation is reduced to an intellectualized process of acquiring knowledge about objects in an environment. Although a limited faculty in Kant's view, touch produces knowledge of an object's form through physical contact with its surfaces and, hence, is "the most important and the most reliably instructive of all senses" (Kant 1978, 41). In association with sight and hearing, touch leads to empirical understanding; touch is an extension of the human subject towards the object. This non-dialogic concept of sense involves touching the world rather than a being touched by; hearing the world rather than being heard; seeing rather than being seen; being against rather than being with. It is predicated on control of the external world as a collection of objects. Furthermore, Kant links sight to light, transcendence, the sublime, and intellection: "The sense of sight, while not more indispensable than the sense of hearing, is, nevertheless, the noblest, since, among all the senses, it is farthest removed from the sense of touch, which is the most limited condition of perception" (Kant 1978, 43). In contrast to the higher senses, taste and smell do not result in cognition of an object because the concept of the object (i.e. its form and color) is obfuscated during the sensing process (e.g. the plant is consumed or its volatile oil penetrates the olfactory glands). For Kant, smell is the lowest sense, associated with stench, filth, and decay. For example, a nauseating smell penetrates the body against one's conscious will; in other words, the uncanniness of stench confounds the Kantian sensory order.

This results in the privileging of the "higher" senses and the denigration of the "lower" senses, and underlies our limited conceptualization of aesthetics. Far from confined to the canons of Continental philosophy, the Kantian hierarchy has produced sweeping effects in a number of disciplines. For example, developmental psychologist Ernest Schachtel's *Metamorphosis: On the Development of Affect, Perception, Attention and Memory* (originally published in 1959), although not referring to aesthetics per se, defines objectification in comparable language to Kant as the process of perceiving objects that exist outside of the human subject (Schachtel 1984, 85). For Schachtel (as for Kant), the higher sense of vision is object-oriented and, therefore, makes possible the cognition of exterior things. Taste, smell, proprioception, and thermal recognition (i.e. touch) are lower, primitive, and "objectless." Schachtel distinguishes between the allocentric (vision and hearing) and the autocentric (gustatory, olfactory, thermal, and proprioceptive) senses. The allocentric senses exhibit a capacity for objectification and are intellectual and spiritual; the autocentric senses lack the powers of objectification and are physical (Schachtel 1984, 89).

Whereas the allocentric senses can be projected at objects over distances through space, autocentric sensations are localized in or near the mediating sense organs. Autocentric and allocentric sense experiences diverge in terms of how they are remembered and communicated in language. The geographer J.D. Porteous takes up Schachtel's allocentric-autocentric dichotomy in the context of environmental aesthetics (Porteous 1989, 1996). Autocentric (subject-centered) perception is concerned with pleasure and feeling; allocentric (object-oriented) perception relates to objectification, cognition, and knowledge-production, in the Kantian mode (Porteous 1996, 31). In sum, the autocentric senses are physical, "primitive," effective at close-range, and acute in children; the allocentric senses are chiefly visual, intellectual, "sophisticated," detached, distanced, easier to recall, and more developed in adults. These sorts of taxonomies limit the importance of the lower senses and have serious implications for the design of urban environments today and, more generally, for coming to terms with our relationship to the natural world in the era of climate change.

The FloraCultures aesthetic framework encompasses the traditional conceptualization of aesthetics as beauty (based predominantly or entirely in sight) and aesthesis as immanence and sensation. I propose the term *floraesthesis* as a place-based model of aesthetics drawing upon the concept of aesthesis as experience gained through the multiple senses. However, most theoretical accounts of aesthetics, perception, and the senses—Adorno, Kant, and Schachtel's included—neglect aesthesis, the etymological and conceptual root of the Western formulation of aesthetics as visual appreciation. In Kant's example, sensation (aesthesis through touch) merely serves object cognition. However, as the derivation of Alexander Baumgarten's 18th-century neologism aesthetics, aesthesis connotes sensation and embodiment—modes of experience cultivated through touch, taste, and smell in dynamic interplay with hearing and vision (Mules 2008). An investigation of aesthesis entails the concerted exploration of the senses independently of hierarchies (lower vs. higher, autocentric vs. allocentric, cognitive vs. non-cognitive). The oppositions promulgated by such aesthetic models fail to produce well-rounded readings of aesthetics and understandings of human embodiment in place—as if autocentric and allocentric perception can be divided; as if taste and smell can operate independent from vision, hearing, and touch; as if the activities of the brain (cognition) can be cleaved from the sensations of the body (aesthesis). Aesthesis is the foundation for exploring the integration of the senses.

Interpreting Botanical Heritage through Floraesthesis: Some Examples from Perth

Floraesthesis is an embodied aesthetics of plants—one summoning the five senses while also inviting the vital senses that exist outside of the five-sense regime, including proprioception (bodily awareness in space) and topaesthesia (sense of place) (Ryan 2012a, Chapter 13). It is also the critical perspective I use to analyze some of the writings from 1829–1929 collected during the FloraCultures research. Through the intensive exploration of multi-sensoriality and sensation, we can appreciate the extent of colonial-era interactions with Perth's plants. Thinking plants, aesthetics, and sustainability together, I note that the historical examples that follow are not single-mindedly attuned to visual appreciation, but also to embodied aesthesis through a variety of activities: the harvesting of plant material for housing, the making of preserves from native fruits, and the appreciation of the endemic perfumes of the bush. Also embedded within the writings are statements about the increasing need to conserve Perth's endemic plant life, even during the early days of the colony.

An embodied aesthetics of marri (*Corymbia callophylla*) is apparent in the accounts of its dark-red resin or kino and the use of its flowers for a beverage. Yet, the tree's scientific name reflects a purely visual aesthetics; the genus *Corymbia* refers to the flower structure while the species name *callophylla* is for "beautiful leaf." In the 1880s, colonist and lawyer George Fletcher Moore recorded *kardan* as the Whadjuck (metropolitan area) Nyoongar name for "Eucalyptus resinifera; red gum-tree, so called from the quantity of gum-resin of a deep congealed blood colour, which exudes during particular months in the year, through the bark" (1884/1978, 28, "A Descriptive Vocabulary"). To the Menang (South Coast) Nyoongar, *marri* defined as "flesh or meat" (Moore 1884/1978, 51) connotes the sensation of eating. Moore lists *numbrid* as "the flower or blossom of the red gum-tree, from which the natives make a favourite beverage by soaking the flowers in water" (1884/1978, 62). For the English novelist D.H Lawrence, who visited Perth briefly in the 1920s, kino symbolized the melancholy of the bush in physical terms: "This tree seems to sweat blood. A hard dark blood of agony. It frightens me—all the bush out beyond stretching away over these hills frightens me, as if dark gods possessed the place" (quoted in Skinner 1972, 112). Despite Lawrence's horror, kino has an extensive history of medicinal use, some of which was of practical interest to colonists. In 1836, the missionary Francis Armstrong (1979, 199) recorded the use of kino amongst the

Nyoongar as an antiseptic for wounds, whereas in the 1880s the homesteader Ethel Hassell (1975, 24) wrote of its use in treating diarrhea amongst settlers. As for the commercial viability of the kino, Moore (2006, 14) observed that "much gum might be collected from the red gum tree. It is said to be a powerful astringent and might be useful in that way or would make a good varnish. I shall try to send you some specimens of it and the white gum from the zanthorrea [*sic*] which here is familiarly called 'black boy'."

During the time period, there was a proliferation of written information about balga, known by the common names grasstree and—formerly—blackboy. Although not related to the Gramineae family, the moniker *grasstree* comes from the frond-like appearance of its foliage. The name *blackboy* reflects the colonial-era perception of its trunk, thought to resemble the distant appearance of an Nyoongar person in the bush. The genus *Xanthorrhoea* derives from the Greek words *xanthos* (yellow) and *rheo* (flow) in reference to its sap. Numerous parts of balga have been consumed traditionally by Aboriginal peoples, making the species the "refrigerator and one-stop shop of the bush." Its leaves (*mindarie*) have been used for thatching huts (*mia-mias*) or making torches, dead flower stems for fire and spear-making, living flower stalks soaked in water to produce a fermented drink, and its aromatic sap combined with kangaroo dung and ash to make an early epoxy. The dark black trunk of balga contains nourishing *bardi* grubs. In the 1880s, Moore recorded the traditional uses of the small but long-lived tree, reflecting an embodied aesthetics of touch and taste in particular:

> This is a useful tree to the natives where it abounds. The frame of their huts is constructed from the tall flowering stems, and the leaves serve for thatch and for a bed. The resinous trunk forms a cheerful blazing fire. The flower-stem yields a gum used for food. The trunk gives a resin for cement, and also, when beginning to decay, furnishes large quantities of marrow-like grubs, which are considered a delicacy. Fire is readily kindled by friction of the dry flower-stems, and the withered leaves furnish a torch. (1978, 3, "A Descriptive Vocabulary")

An embodied aesthetics or floraesthesis of balga was also at play during the colonial era, during which some of its Nyoongar uses were adopted. In 1927, Mrs. Edward Shenton published her recollections of her mother's account of arriving in Perth in 1830. Shenton begins with a description of the city's flora, as related by her mother in the aesthetic language of scenery, beauty, and pleasing coloration. The natural attractiveness of the soon-to-be-urbanized landscape persisted in Mrs. F. Lochee's memory:

> My mother [nee Emma Purkis] said she could never forget the beauty of the scenery when they arrived in Perth. From Point Lewis to Mill-street, she said, there was a high hill running right down to the river and the bank as far back as Hay-street was very beautiful in colours of green, yellow, white and pink, with small streams running at intervals into the river. From Mill-street commenced another long hill, which ran as far as Bennett-street, and was clothed in beautiful verdure and broken

by gaps and running streams. Where Government House ballroom now stands was a ravine and a running stream. (Shenton 1927, 1-2)

Shenton's recorded oral history speaks of the effects of colonization on the bushland in an area now quite close to Perth's Central Business District where the Purkis family accepted a land grant: "When they arrived the bush had been cut to make St. George's Terrace the same width as now from Milligan-street to Lord-street (now Victoria-avenue), and there was a narrower cutting (Adelaide-terrace) to a decline in the roadway" (Shenton 1927, 2). Although they arrived in Perth without their belongings, the family soon had accommodation comprising "one large tent with a fly, which was divided into three rooms for sleeping, lined with blankets and carpeted with blackboy rushes over which mats were placed" (Shenton 1927, 2). The first part of the quoted text from Shenton perfectly exhibits a visual aesthetics of flora, with its Kantian emphasis on "the beauty of the scenery," "colours of green, yellow, white and pink," and the Perth landscape "clothed in beautiful verdure." However, Shenton's description of the living arrangements of her mother's family suggests a different kind of perception: a floraesthesis of balga, its "rushes" or mindarie foliage supplying physical comfort, the sensation of warmth and the pleasure of its organic fragrance in their new home.

Like balga, zamia provided more than visual aesthetics to Swan River colonists (see Chapter 3). As I discussed earlier in *Being With* in relation to plant narratives, zamia (*Macrozamia riedlei*) is a member of the Zamiaceae family of cycads distributed throughout Australia and prominent throughout Kings Park. Zamia fruits were detoxified by Nyoongar people through a variety of fermentation processes, roasted, ground, and baked into a nutritious damper. Moore (1884/1978, 17) recorded the fruits as *by-yu* and observed: "This in its natural state is poisonous; but the natives, who are very fond if it, deprive it of its injurious qualities by soaking it in water for a few days, and then burying it in sand, where it is left until nearly dry, and is then fit to eat." Its "wool" was used for clothing, bedding and children's diapers in the early years of the colony, as the following remark in *The Western Mail* indicates: "The wool which surrounds the bases of the leaves has uses for stuffing pillows, cushions, etc." (The Western Mail 1922, March 2, 30).

The consumption of zamia fruits or the soaking of balga flowers to produce a fermented drink are examples of floraesthesis in which a plant substance is taken into the body. Nevertheless, despite these examples from the historical record, Perth's flowering shrubs have often been constructed in the exclusively aesthetic language of form and color. For example, an article in *The West Australian* lists the popular names of the endemic *Hypocalymma robustum* as Swan River myrtle, pink myrtle, pink-all-the-way-up, monkey blossom, wild peach, and pink heath, and goes on to say: "Its slender branches are erect and rigid, giving the effect in springtime of thick clumps of graceful spikes, closely set with vivid pink blossom fully opened and with innumerable yellow stamens at the base of the stems" (The West Australian 1928, September 7, 7). A plea for conservation follows its aesthetic depiction: "All who remember how plentiful this lovely plant once was around Perth regret its rapid disappearance in recent years from the open spaces in the vicinity of the metropolis" (The West Australian 1928, September 7, 7).

Stinkwood (*Jacksonia sternbergiana*), however, evokes noxious smell rather than visual pleasure, "so called because the wood when burned gives off evil-smelling gases" (The Western Mail 1929, February 7, 38). Moore (1884/1978, 40) described the species as "one of the dullest and most melancholy foliaged trees in Australia. It has an unpleasant smell in burning, from which it is frequently called stinkwood." Whereas Swan River Myrtle impressed the eyes and stinkwood repulsed the nose, broom ballart (*Exocarpus sparteus*) pleased the taste buds, but only somewhat for Moore (1884/1978, 24) who described the berries as having "no particularly good flavour."

Like the marri, balga, and zamia, many of the herbaceous plants in the bushland of Kings Park bear cultural histories involving the human appreciation of plants through the "lower" senses of taste, touch, and smell. For example, bloodroot (*Haemodorum spicatum* or *bohn* in Nyoongar) derives from *haima* for "blood" and *doron* for "gift" (as with the kangaroo paws discussed in Chapter 4). Moore recorded the flavor of the root as resembling "a very mild onion. It is found at all periods of the year in sandy soils, and forms a principal article of food among the natives. They eat it either raw or roasted" (Moore 1884/1978, 12, "A Descriptive Vocabulary"). Accounts of milkmaids (*Burchardia congesta*) evoke the visual and haptic senses and, specifically, the sensation of touching wax. Perth children were known to collect the flowers: "The flowers of Burchardia are very beautiful individually. The yellow stamens stand upright on the points of an inner white star with a dark-coloured, reddish-green, conical seed-capsule [...] it is commonly called 'wax' by children, who love to gather the snowy blooms" (The West Australian 1924, September 26, 6). Another herb, *Stirlingia latifolia*, is known as blueboy or rust flower. The name *blueboy* refers to the fact that wall plaster—made with sand from where the species grows—turns blue. In 1918, essential oil of the species examined at the Imperial Institute in the United Kingdom was found to consist almost entirely of acetophenone, which has soporific properties and was researched for its medicinal value during the period.

Conclusion: Toward Aesthesis

Approached from the perspectives of sustainability and aesthetics, these examples collectively suggest the importance of Perth's indigenous plant life for its early European inhabitants. In addition to providing insights from the period following settlement, the writings also reveal aesthetic and conservation attitudes toward flora. The perceptions represented in the written materials neither reflect sustainable practices adopted by the settlers toward the Perth environment nor show an ethical regard for flora beyond the use-value paradigm. They do, however, indicate a wider non-scientific, cultural appreciation of and concern for Perth's environment that I maintain has been largely diluted in the public population of the metropolitan region today. While some aesthetic perceptions are exclusively visual, others reveal embodied interactions with an abundant, diverse, and compromised flora. The greater physical distribution of New World plants in the fledgling city and the lack in the availability of essential goods (e.g. food, medicine, fiber, ornamentation, and building materials) in the early years of the State necessitated a gamut of embodied interactions with flora. Such human-plant transactions engaged the "lower," intimate senses of touch, taste, and smell—which have been atrophied in the Western model of nature aesthetics but which are essential to engendering long-term sustainable relationships with the non-cultivated botanical world.

In developing an analytical approach to FloraCultures, I have found it necessary to rethink the origins and applications of environmental aesthetics and the role of the senses. I have set out to re-read the cultural heritage of plants with attention to aesthesis—not constrained by visual or cognitive bias. A phenomenological aesthetics co-constitutes subjects and objects in a world in which "objects" (here, plants) are regarded as sentient, sensuous, autonomous, and agentic. There are lessons in such philosophy and history for today's design practitioners. Sustainability research should be aware of the history of place and should consider the autocentric and allocentric senses as two sides of a coin. Sensuous readings of human-plant histories can help the designers, planners, conservationists, ecologists, architects, and educators of today create the pleasurable metropolitan landscapes of tomorrow. These designed places would not only call attention to the beauty of wildflowers, but would simultaneously encourage physical interaction with plants as co-inhabitants of our suburban neighborhoods and with associated benefits to animals, fungi, insects, and other living beings. A radical and subversive form of metropolitan gardening would put edible native plants amongst ornamentals— food and beverage at the peripheries of lawns and playing fields—while specifically protecting endemic species that are endangered or threatened. Our sense of place and plants is an embodied one— a reality also expressed by poets of the region.

Chapter 6: "The Name Blossomed": Landscapes, Habitats, and Botanical Poetry

Introduction

Ecopoetry has the potential to narrate human experience in which the senses engage with ecological processes and environmental phenomena. Indeed, physical interaction with nature is a pivotal but under-scrutinized dimension of ecopoetry. In this chapter, I will explore these assertions by characterizing sense-rich, ecologically networked poetry as *habitat poetry*. In addition to sensory-fullness, another defining quality of habitat poetry is its representation of the lives of people, plants, animals, and fungi within their broader ecological interrelationships: *being with* the natural world *in* and *through* poetry. Moreover, habitat poetry also tends to convey a poet's grappling with scientific discourses. These three markers of habitat poetry (*ecology*, *sense*, and *science*) will be articulated in the regional context of the South-West. The works of South-West poets Alec Choate (1915–2010), Andrew Lansdown (1954–), and John Kinsella (1963–) make use of sensory language that expresses each poets' experiences of landscape, while conveying something about the dynamics between science and poetry (see also Chapters 2 and 3). The works of poets Glen Phillips, Dorothy Hewett, and Tracy Ryan evoke the oscillation between domestic and ecological settings. The *habitat* concept furnishes an interpretative framework for reading these six poets as not only landscape poets but, more precisely, habitat poets. This distinction is pursued in this chapter through the theoretical positions of Cosgrove, Elliott, and Giblett in particular. Whereas landscape poetry tends to prioritize visual experiences (and thus, visual aesthetics, as discussed in Chapter 5), habitat poetry demonstrates human engagement with the natural world through sensory plurality and an acute awareness of ecology and science. Focusing on these six South-West poets, this chapter calls attention to poetic works that address flora through broad ecological understandings, or what will be referred to as *habitat awareness*.

Landscape Poetry and Habitat Poetry

Unlike animals that move, gaze, and vocalize, plants are commonly perceived as static objects or two-dimensional surfaces (Ryan 2013, Chapter 6). Geographers Russell Hitchings and Verity Jones observe this cultural tendency in arguing that "vegetation is something passive in contemporary understanding: to be in a vegetative state is to be without mind. Yet the root of the word 'vegetative' [etymologically] is associated with activity and enlivened animation" (Hitchings and Jones 2004, 11). Published in 1927, *The Lure of the Golden West* by Thomas Sidney Groser demonstrates the common perceptual tendency to appreciate the flowering plants of the South-West for their visual and vegetative qualities. The following excerpt from Groser exemplifies the language governing visual appreciation with such phrases as "lovely picture," "prevailing colour," and "glowing pageantry." Because it is prose and not poetry per se, the rhetoric of ocularcentrism with regard to plants is plainer and more evident:

> There is scarcely a more lovely [*sic*] picture imaginable than a West Australian Bush in the Springtime. Pink is perhaps the prevailing colour—certainly where the 'everlasting' predominates [...] The rich green undergrowth of Spring-time [*sic*], and the evergreen and flowering eucalyptus trees, form a rich setting for this glowing pageantry of colour. (Groser 1927, 216)

The speaker emphasizes the springtime color and form of flowers. The represented scene implies the two-dimensional fixity of the observed plants and their separation from the pollinators or landforms co-occupying their space. Groser's extract treats flora—lovingly and appreciatively still—as objects of art. This writing—which I will refer to as *landscape*—focuses on the visual impact of flowers, privileging qualities possessed by a plant only at certain times of the year, namely the beauty of flowers during particular seasons (Chapter 1).

Groser's language has its roots in a landscape tradition in which sight alone is enough to generate spiritual, emotional, and aesthetic experiences of nature. The Romanticist subject, informed by a highly ocular modes of appreciation, tends to construct nature as a vista, prospect, or vantage point—in other words, a landscape—through the primacy of the sense of sight (Giblett 2011, 68-72). John Barrell argues that the word *landscape* was introduced into English "from the Dutch in the sixteenth century to describe a pictorial representation of the countryside" (Barrell 1972, 1). Nature writing in English owes its origins to the works of landscape poets such as Wordsworth whose early writings in particular reflect an ambivalent relationship to "the bodily eye [...] the most despotic of our senses" (Wordsworth quoted in McKusick 2000, 56). Romanticist poetics often privilege landscapes of various aesthetic modalities—the sublimity of mountainscapes, the picturesqueness of valleys, or the beauty of well-formed natural objects, including flowers: "But lately, one rough day, this Flower I passed | And recognized it, though an altered Form | Now standing forth an offering to the Blast | And buffetted at will by Rain and Storm" (Wordsworth 1803, "The Small Celandine"). Although Wordsworth's celandine exhibits some agency in "standing forth," the flower is constructed simply as one of nature's forms, impacted upon by the elements, rather than a plant-as-subject producing physical sensations that affect the poet intimately through touch or smell.

Differently put, landscape poetry involves a way of seeing the natural world that reflects certain rhetorical and aesthetic conventions of sight and language (for further background, see Berleant 2005, Cosgrove 1998, Elliott 1967, Heidegger 2009). Brian Elliott in *The Landscape of Australian Poetry* defines landscape as "the visible scene about us, the subject-matter of descriptive picture-making" and confesses that when seeking synonyms "I have sometimes employed the term *topography*, or referred to the *vista*, signifying what is seen by the eye [italics in the original]" (1967, xi). In a similar fashion, the geographer Denis Cosgrove understands landscape as "not merely the world we see, it is a construction, a composition of that world. Landscape is a way of seeing the world" (1998, 13). A way of seeing implies modes of viewership; the resulting ocular (or ocularcentric) language is of beautiful flowers, sublime forests, or picturesque heath lands—a connection between discourses and landscapes explored by Arnold Berleant in *Aesthetics and Environment*. Extending Cosgrove's notion, landscape poetry could be theorized as a mode of constructing the world in imagistic terms (see also Heidegger 2009, 207-223).

A conceptual distinction between *landscape* and *habitat* allows various aspects of human engagement with the botanical world to be identified and better understood in terms of environmental writing. Is the poem set back distantly, painting a "lovely picture"? Or does it communicate intimate sensory contact with plants through embodied acts of tasting, touching, smelling, listening attentively, and looking carefully? The work of cultural theorist Rod Giblett in *People and Places of Nature and Culture* provides a structure for teasing out the points of difference between landscape and habitat poetry. According to Giblett, landscape poetry inherits the problematic aspects of the *landscape* concept (2011, 66-68). In particular, the categories of the beautiful, sublime, and picturesque prioritize the visual perception of nature's surfaces and the formation of value judgments based on limited sensory information (Giblett 2011, 57-94). *Landscape* implies, most commonly, an externalized scene or visually demarcated object. However, here Giblett differentiates between writing that constructs *landscape* through vision almost exclusively—propounding a Kantian hierarchy of the senses—and writing that engages the complexities of nature through diverse (and often messy and unruly) sensory entanglements:

> Landscape writing that aestheticises the static surfaces of nature can be contrasted with nature writing that celebrates its dynamic depths. I define nature writing as the creative, written tracing of the bodily and sensory enjoyment of both the processes and places of nature. (2011, 26)

Whether through the creative intent of the poet or not, environmental writing that emphasizes the surface features of plants reflects a landscape mode of perception, according to Giblett, Cosgrove, and other environmental thinkers.

Conversely, *habitat* refers to qualities of ecological interdependency and sensory plurality in writing. This is not to say that these qualities will always absent in the mediation and experience of landscape; however, sensory plurality can be constrained by the privileging of vision (for background on ocularcentrism, see Crary 1990, Jay 1993, Levin 1993). The term *habitat* is derived from a Latin verb, "to live, dwell," later developing into the noun form, "dwelling place" (Harper 2012a). Sense-rich writing about flora—entailing embodied human experience of plants—could be described more broadly as *nature writing*, a literary form that engages the processes, patterns, and sensations of a place and its inhabitants (Ryan 2013). As a genre of nature writing, habitat poetry tends to allude to plants in close relation to fauna, rocks, water, and other natural and cultural features that constitute the land. As it will be propounded in this chapter, habitat poetry comprises three interweaving characteristics: ecology, embodiment, and science. Communicating the network in which the plant is situated as a subject rather than an object of research (in particular, see Latour 1999), habitat poetry enlivens a reader's awareness of biological rhythms and the interdependencies between species.

In many of the following works by South-West poets, a general curiosity about a plant in its habitat leads to an embodied investigation: smelling, touching, tasting, listening, and looking carefully, as well as experiences of synaesthesia between the sense faculties. In the examples from Choate, Lansdown, and Kinsella given in the next section, *landscape* and *habitat* modes work hand-in-hand. Ecological, bodily, visual, and scientific interests in the plant remove the downfalls of detached visual spectatorship. For the nineteenth-century American prose writer Henry David Thoreau, who is credited with the emergence of nature writing in North America, botanical habitats (such as swamps and forests) offered comparable participatory immersion in ecology through the senses (Chapter 3). In this regard, Thoreau wrote of habitats rather than landscapes exclusively. His sensory entanglement with botanical nature countered the physical detachment often at the center of the landscape mode, as the previously cited example from Groser makes evident.

Reading Lansdown, Choate, and Kinsella as Habitat Poets

The works of Alec Choate, Andrew Lansdown, and John Kinsella convey habitat awareness of South-West plant ecologies. In sum, their poetry is about process, sensation, and science. As these extracts will intimate in different ways, their work engages directly with botanical science through references to taxonomic names and anatomical terms. Simply put, plants evolve in these poems, and are represented as part of a dynamic, changeable nature. Expressing habitat awareness, human perception synchronizes in their poetry to the *poiēsis* of plants over time and the seasons (see Chapter 1). Their writings include visual and non-visual experiences—although not always both—as a way of representing the multidimensionality of plants and human sensory experience of the botanical world.

Andrew Lansdown (1954–) was born in Pingelly, a regional town in Western Australia's Wheatbelt. His first published collection, *Homecoming* (1979), clearly shows his interest in Perth-area flora and fauna from the perspective of close sensory contact. *The Oxford Companion to Twentieth-century Poetry in English* characterizes Lansdown as "a miniaturist, a poet attentive to the smallest details of nature" (Grant 1994). Dennis Haskell and Hilary Fraser describe Lansdown's work as "reminiscent of Wordsworth and Coleridge" (Haskell and Fraser 1989b,

122), although I suggest that Lansdown is more sensorial and embodied than the Romantic label can afford. His recent publications, including *Birds in Mind: Australian Nature Poems* (2009), further round out Lansdown's thirty-year identity as a nature writer interested in the senses. "A Few Weeks Later I Return To Find" from *Homecoming* (1979) exemplifies his sensory curiosity about South-West flora. In this poem, visual appreciation parallels the ecological processes of the balga flower (Figure 6). Tactile interaction between plant and poet creates palpable and intimate language. Lansdown moves between bodily encounter and visual analysis of the balga's morphology and symmetry. Moments of sensory contact interweave with a kind of taxonomic itemizing of the flower:

> Centred in the stamens,
> the shorter stylus – surrounded,
> and at times, 'covered
> by a glistening glob of transparent nectar
> which, in turn, was caught in the cup-pit
> of the six guardian stamens. (ll. 11-16)

After an anatomical inventory of the flower's stylus and stamens, Lansdown engages his other sense faculties. The poem tracks up and down the balga stalk, indicating an ecological awareness of this particular *Xanthorrhea* as a mutable individual and member of its species. Playful curiosity infuses straightforward morphological descriptions. Throughout, Lansdown employs bodily tropes like "tiny yellow vulvas of pollen" (l. 10). Inquisitiveness and sensory openness reach an apotheosis when:

> I thought each flower had mysteriously
> caught last night's dew,
> so I put my tongue to it
> (Descartes would not have approved) to see:
> it was a powerful, honey-thick
> nectar. The odour was a heavy
> sweetness. I wiped the pollen from my nose. (ll. 17-23)

Lansdown's poem expresses a light defiance of scientific knowledge construction and objective methods of knowing—"Descartes would not have approved" (l. 20). Such sense acts could be read as deconstructing scientific epistemologies. Scandalous intimacy with the balga calls into question the very notion of objective detachment: "so I put my tongue to it" (l. 19). The poet's sensuous body as a whole participates *in*—rather than distantly views—the visual manifestations of the balga's ecological changes. Vision—"to see" (l. 20)—links to taste—"a powerful, honey-thick | nectar" (ll. 21-22)—and smell—"The odour was a heavy | sweetness" (ll. 22-23). The poet's response is embodied: "I wiped the pollen from my nose" (l. 23).

Opposing qualities of touch tend to characterize the human experience of many species of South-West flora (Seddon 2005). For example, several species of dryandra bear down-soft flowers surrounded by stiff and thorny foliage (Collins, Collins, and George 2008). Lansdown demonstrates that the balga offers a range of sensation to human (aesthetic) subjects. After using its colloquial name "blackboy" in a direct address of the plant, Lansdown notes subtleties in how the flower stalk feels:

> Blackboy,
> the compact, coarse sandpaper
> of your flower-spear
> has turned to softness. (ll. 24-27).

Judging from the phrase "flower spear," Lansdown appears to acknowledge the balga's capacity for *autopoiesis*—self-generated evolution over time as manifested by coarseness and armour, softness and suppleness. The poem's title similarly intimates an awareness resulting from the direct experience of plants (or the same plant) at different times through the seasons. The balga *in situ* is not a fixed object of perception but rather a highly dynamic phenomenon existing in ecological space. Awareness of the balga's habitat and of the balga *as* habitat is evident in the last stanza:

> Later
> the seeds will come;
> then the parrots.
> I see the sharp, triangular lengths
> of your deep-green leaves
> shimmer in anticipation. (ll. 28-33)

Figure 6. This is a detail of a typical *Xanthorrhoea* (balga) flower. Visual images of these species could contradict a position for heightened sensory experience of plants through tasting, touching, smelling, and listening closely, rather than through the pleasure of sight alone. However, my hope is that these photographs become a point-of-reference for the readers outside of Western Australia who may be unfamiliar with the plants poeticized by the writers included in this chapter.
Source: FearTec, Wikimedia Commons (GFDL permission).

Instead of extracted from its ecological setting and appreciated as a stand-alone tree (if that were even possible), the balga in "A Few Weeks Later" is contextualized ecologically: seeds transform into different forms and parrots come and go. There is both an *imminence* and *immanence* of plants in Lansdown's work.

Endemic South-West plants also figure conspicuously in the works of Alec Choate (1915–2010). Born in Hertfordshire, England, Choate spent most of his life in Western Australia. Haskell and Fraser observe that "his poetry is distinguished by its extraordinary descriptive power, as well as by its ability to suggest, through metaphor, a mysterious chord of sympathy between nature and humanity" (Haskell and Fraser 1989a, 34). Much of Choate's poetry deals with the desert regions of the State where he worked as a land surveyor (The University of Western Australia 2010). The collections *Gifts Upon the Water*, *A Marking of Fire*, and *Mind in Need of a Desert* exhibit Choate's ongoing fascination with the ecologies of dryland species. The poems "Land in Flower" (1986, 14) and "Prison Tree, Derby" (1978, 35) represent his interest in the linkage between botanical ecologies and cultural histories. However, "Nuytsia Floribunda" and "Poverty Bush" particularly demonstrate Choate's characteristic interweaving of botanical science, taxonomic contemplation, and poetic sensibility in the ecological context of WA's arid country.

"Nuytsia Floribunda" is the scientific name for the endemic West Australian Christmas Tree (Choate 1986, 16) (also see Chapter 3 and Figure 7). The infusion of science into poetry is reflected by Choate's rhetorical decision to use the taxonomic name as the title, rather than the plant's more poetic colloquial names *cabbage tree* or *Fire Tree* (for a historical discussion of the species, see Lindley 1840). The poet considers the tree in its habitat, while refraining from the use of technical terms such as *hemi-parasite* or *haustoria* common to botanical accounts of *Nuytsia* ecology and physiology (see, for example, Hopper 2010). For Choate, *Nuytsia* is symbolically, cultural, and ecologically vital. For instance, the tree in flower is an emblem of the South-West Australian summer:

> This tree could only find root
> in a land whose heart belongs
> to its summer, no other
>
> season, and where the summer
> feels bound to repay that heart
> with an emblem of grandeur. (ll. 1-6)

The middle stanzas trace the progression of the seasons and set the context for the flower's arrival as an "emblem of grandeur" and a harbinger of the hot season:

> Autumn and winter linger
> to make ready as tinder
> all that is dark in their time,
>
> the dark of life waste and mould
> and rain when blind among roots,

the spleen of night's halting hours

or simply the crape of clouds.
Spring is the easing of limbs,
the young leaves winking for warmth. (ll. 7-15)

Figure 7. This is a detail of mudja (*Nuytsia floribunda*) in flower in November 2010 in Dianella, WA. Following heavy spring rains, 2013 was also an outstanding year for the flowering *Nuytsia*, although urban populations of the species continue to be cleared for development around Perth Airport.
Source: Author.

In later stanzas, the fire and flower become analogues for the colonial trope of *Nuytsia* as a "tree on fire" (Lindley 1840), where the tree's flower signifies the beginning of the bushfire season:

But when the summer returns,
the culminating summer,
it breathes upon the tinder,

and from the gauntlets of green
of this chosen tree in its
thousands, lights torch after torch

of amber that floods to gold,
its own heart naked as fire
the land's heart naked as flower. (ll.16-24)

As homophones, fire and flower represent the enduring regenerative processes of the land (for a classic account, see Hallam 1975). Described as "amber that floods to gold" (l. 22), flower color alchemizes as the season progresses. Rather than an eschatological end, the golden blossom is a culmination—or even crescendo—of a string of interactions between "waste," "mould," "rain," and "roots." Here, the flowering *Nuytsia* is not represented as isolated from its habitat. Instead, the beautiful blossom is positioned within a broader network of ecological agents—a habitat. By the poem's end, fire is no longer a mere trope or metaphor; it refers to the material reality into which *Nuytsia* projects upward during the wildflower and wildfire season. The poetic effect shuns the representation of the tree as a passive object of study; metaphorically and ecologically, fire infuses flower, flower fans fire, and the tree's being is revealed in these terms:

> while from skyline to skyline
> the haze is a secret's veil
> that a fierce trust has shredded. (ll. 25-27)

The "fierce trust" shredding the haze of smoke is the uplifted, golden canopy of *Nuytsia*. The flower is intrinsically a part of the habitat, a symbol for the regeneration of the land from season to season through autopoietic processes. In this sense, the poem represents plant ecologies imaginatively through the metaphor of a pact between humans, plants, animals, landforms, and the elements fire, wind, and water. Metaphysically, trust (as a quality of autopoiesis) makes possible the cycle of decay and growth, flowering and firing. Choate's layered reading of the *Nuytsia* flower contrasts strikingly to surface-oriented appreciation of flora, theorized earlier in this chapter.

In "Poverty Bush," Choate (1995, 94-95) comments on the historical misperceptions inscribed in botanical names vis-á-vis the desert plant *Eremophila alternifolia*. Poverty bush is concentrated on the arid eastern edge of the South-West. *Eremophila* is from the Greek roots *phila* for "to love" and *eremo* for "lonely places" or "desert." The poem opens with two tercets that affiliate the plant with love for the desert, as its scientific name indicates. However, the love is tainted by thwarted hope, as its common name denotes:

> The desert cries out for love,
>   and no shrub answers
> With more heart to return it
>
> than this whose vast sisterhood
>   is far from the name
> and seeming of poverty. (ll. 1-6)

Poverty bush is perfectly adapted to desert country, but the same soils in which it prospers have a history of impeding colonial pastoral intrusions. Choate historicizes the common name, characterizing it as the miscue of European settlers:

> Poverty of mind rather
>   was theirs who crammed mouths
> on the pastoral reaches
>
> and who when the ground feed died
>   so named the shrub, it
> being no browser's standby. (ll. 7-12)

With tough and spiky foliage, poverty bush is resistant to grazing by animals. On the one hand, plant names are linked to the conversion of Indigenous land. On the other, the metaphorical nuances of scientific names reveal the plant's affinity for desert ecology and its adaptation to dry conditions. Toward the end of the poem, the act of naming involves the messianic arrival of a botanist who redeems the plant from cultural misrepresentation:

> But someone came, someone saw
>   it lacquer its leaves
> against the wind's rainless lips,
>
> scatter its seeds and trust roots
>   to the rust-red sand,
> saw how it decored itself
>
> in wool, a ripple of scales,
>   a mantle of hair,
> or posed sepals as petals,
>
> its means, and its miracles,
>   for coming to terms
> with skies and their gaze of stone.
>
> Someone came. The name blossomed.
>   '*Eremophila*,'
> he said, 'or Desert Loving'. (ll. 13-27)

The plant's technical name calls attention to the ways in which the species has adapted to the same conditions that repelled early settlers. This skillful poeticizing of environmental history reveals Choate's familiarity with ecological processes in the phrases "scatter its seeds" and "posed sepals as petals." The habitat is austere. "The wind's rainless lips" evokes the parched earth, while "skies and their gaze of stone" evokes the omnipresence of the desert sun. The final tercet suggests that scientific names tend to have poetic hues, thus animating taxonomic knowledge with imaginative and figurative dimensions.

    Poetic attention to South-West plants also figures into the poetic works of John Kinsella (1963–), a prominent Western Australian writer and critic who lives most of the year near Northam in the Wheatbelt east of Perth. His poetry displays an acute visual awareness of place, culture, history, and ecology. Unlike Lansdown and Choate, Kinsella is known for upsetting the notion of landscape as a pastoral idyll; indeed Kinsella's landscape is a deeply fragmented yet beautiful one, a polarized habitat for people, flora, and fauna. Kinsella's poetics has been self-characterized as "poison pastoralism" or "anti-pastoralism," the latter term coming from ecocritic Terry Gifford to refer to the tension of "how to find a voice that does not lose sight of authentic connectedness with nature in the process of exposing the language of the idyll" (Gifford 1995, 55). In contrast to a Wordsworthian sense of aesthetic harmony, Kinsella focuses on the immediacy of sensoriality, knowledge of ecology, and linguistic disintegration (see Haskell 2000 for an analysis of Kinsella's pastoral poetics). Kinsella's habitat poetry is rooted in the regional ecological crisis of the Wheatbelt and is not limited to speculative distance and visual appreciation alone.

The slippage between the construction of a landscape as an image and the appreciation of land as a moving sensation is evident in "Everlastings" (Kinsella 1997, 29-30). The poem opens with bodily references used to describe a family gathering wildflowers:

> A couple pick flowers
> while their child lies
> cribbed in the dry rustle
> of stalks & petals. (ll. 1-4).

While a tactile experience, the harvesting of wildflowers is now considered an ecologically unsustainable way of engaging with indigenous plants (see Summers 2011). In the poem, the habitat protects and nurtures the family, as the flower stalks and petals cradle a small child (see Figure 8). Bodily tropes connect people to plants; for example, the everlastings are "broken-necked." Similarly, the short poem "Paperbarks" (Kinsella 1997, 174) uses corporeal language: "skins peel and flake | about the grasping roots" (ll. 6-7). The exfoliated exterior is comparable to a skin peeling back and is the most distinguishing visual characteristic of the paperbark: "absorbent skins will not extinguish when voice | falls and memory lingers, for these are ghosts" (ll. 9-10).

Although human experience of everlastings is not exactly the preoccupation of the poem, "Everlastings" does employ physical imagery to construct a scene that reverberates with sensation. The "painting" (l. 17) of the flowers comes to life as the breeze animates "rippling waters" of everlastings:

> The breeze stirs the feeling
> deep inside the painting,
> the sun flickers
> & is passed over
> by pumice clouds, bunches
> hanging rigid in the shade,
> mock-glorious in their brilliance. (ll. 16-22)

Kinsella's lyricism constructs the environmental setting as a fluid and changing milieu rather than rigid and flat scene. The visual qualities of the everlastings resonate as "White pink rose scarlet" (l. 9). More than ocular and two-dimensional, the colors are synaesthesiac and exude movement: "cool almost ice their swayings" (l. 10). Additionally, the allusions "Bees, laden" (l. 12) and "the sun flickers | & is passed over | by pumice clouds" (l. 18-20) disclose Kinsella's habitat awareness: the everlastings situated in their ecological network rather than represented as aesthetic objects. Similarly, "The Bottlebrush Flowers" (Kinsella 1997, 174-175) augments visual denominations—"bristling firelick" (l. 11) and "a spiral of Southern Lights" (ll. 11-12)—through allusions to the inextricable embeddedness of the bottlebrush in its habitat: "I've also seen | honey-eaters bob upside down | and unpick its light in seconds" (ll. 12-14) (see Figure 9).

Figure 8. This field of pink and white everlastings near Lesueur National Park in Eneabba, WA, represents the appealing picturesque beauty of these flowers during the spring. Despite their ubiquity in wildflower tourism media, everlasting scenes like this are quite rare in most parts of the State.
Source: Author.

Figure 9. This flowering bottlebrush in Cervantes, WA, is representative of most backyard plantings of the species. While providing habitat and food for birds, its dense form also offers welcome shade for human homes.
Source: Gabriele Delhey, Wikimedia Commons (GFDL permission).

"Exposing the *Rhizanthella gardneri* Orchid" (Kinsella 1997, 227) acutely represents Kinsella's habitat poetry. The poem's multisensoriality allows a reader to connect poetic language to orchid ecology. Here, Kinsella's language evolves in relation to the senses, revealing progressively the uncanny ecological events in the underground orchid's biological cycle. The opening stanza describes the orchid's symbiotic associations. The species requires biological mutualisms with other plants; without human intervention, the flower would never be exposed to sunlight and would live out its entire life underground (Brown et al. 2003). In order to survive without the capacity for photosynthesis and perpetually in the dark, like a fungus (see Chapter 7), the unusual orchid depends on the roots of a host tree:

> Above the roots
> of a Broom Honey Myrtle
> the beak of an orchid
> tastes the acrid air.
> Its mouth sweet with flowers.
> Termites roaming the pollen. (ll. 1-6).

Kinsella's word choice is based on gustatory sensations: "the beak of an orchid | tastes the acrid air" (ll. 3-4) and "its mouth sweet with flowers" (l. 5). The poet observes the anatomical parts of the orchid, in relation to its ecological processes, such as pollen dispersion by termites. In the language of botanical science, Kinsella outlines the mutualistic relationship between the orchid, the host tree, and the mediating fungus:

> Saprophyte,
> and guest-host
> to a root-invading fungus,
> its liaisons go unnoticed
> as the scrub is peeled back,
> and are only half-revealed
> with the lifting
> of the surface. (ll. 7-14).

The unnoticed "liaisons" (l. 10) refer to the time-worn transactions occurring beneath the "surface:"

> Excavated,
> its leaves unfold
> and termites roam the pollen,
> its dark heart
> reddening
> with exposure. (ll. 15-20).

The evocation of motion infuses images of the orchid, as "its leaves unfold" (l. 16). The flower reddens while termites—one of the few known pollinators of the orchid—distribute pollen through their frenetic movements. The poem oscillates between the above-ground act of unearthing the flower by the human (aesthetic) subject and the typically buried state of the orchid. Words form at the margin (ecotone) of the *superficial* and the *subterranean*. A Romantic preoccupation with impressive vistas (the sublime), broad-sweeping expanses (the picturesque), or pleasing objects (the beautiful) is clearly not a characteristic of Kinsella's poison pastoralism. Instead, *Rhizanthella gardneri* is vested with ecological intricacies, human embodiment, and the sensory receptivity of the poet as participant in nature.

Domestic Spheres and Natural Habitats

Through examples from Lansdown, Choate, and Kinsella, the previous section asked: How do poets engage with plant diversity through the fusion of ecology, sense, and science? How do writers express embodied interactions with plants through poetic language? How can we read these three poets as writers of habitats who bring the tutored eye of a naturalist to the craft of poetry? The botanical poetry of the South-West featured in this chapter has been analyzed in terms of *habitat* as an expression of environmental writing. The distinction I draw between *landscape* and *habitat* calls attention to an embodied aesthetics of flora (Chapter 5) and ecosystem dynamics, in the latter term. Through sense-rich engagement with plant ecologies, the selected works demonstrate a poetic probing of the depths of the botanical diversity of the region. Furthering the exploration of botanical poetry, this section focuses briefly on the works of Dorothy Hewett (1923–2002), Glen Phillips (1936–), and Tracy Ryan (1964–). Unlike the examples from Lansdown, Choate, and Kinsella, the works of each of these three poets blend ecological observations with ecological and domestic settings. Their poetry exemplifies an interweaving between domesticated spaces—a playroom, a lawn, and "The Farm"—and environmental spaces and natural phenomena. As such, their works collectively provide a different lens for thinking about the concept of *habitat poetry*.

Along with brown boronia (*Boronia heterophylla*), Australian sandalwood (*Santalum spicatum*) produces one of the most evocative, recognizable, and pleasing fragrances of the endemic South-West flora. The poem "Sandalwood" by Perth-born poet and playwright Dorothy Hewett (2001, 66) describes the aromatic effect of the burning wood. Sandalwood's piquant perfume infuses Hewett's recollection of childhood—one in which the decline of sandalwood and loss of her father are interrelated. The poem opens with a reference to the clearing of sandalwood by cutters in the nineteenth century:

> Our father brings in the last stick of sandalwood
> to lay it reverently on the playroom fire
> the sandalwood cutters moved through this country
> systematically cut it out a generation ago. (ll. 1-4)

"The last stick of sandalwood" is an object of mourning. The momentum of the poem, however, quickly turns to the mysterious drift of scent, the dancing of flames, and the bending figure of a man. The smoldering sandalwood permeates the domestic space of her childhood home with a scented arabesque:

> The heavy scent fills the room the flame dances
> yellow and blue and green on the fluted clock
> till our eyes glaze over our father's dark face
> bending to tend the fire ... (ll. 5-8)

The colorful gyrating flame contrasts to the image of her father "bending to tend the fire," accentuating the tender interaction between his body and the stick of sandalwood. The aroma transforms the playroom into a numinous space and sheltering domain—an "Aladdin's cave." At the symbolic shift when "in a moment everything changes," the burning incense moves beyond its domestic quarters and becomes absorbed back into its ecological origins:

> and in a moment everything changes
> the playroom burns through the night
> like Aladdin's cave it skims through the doors
> floats high clearing the creekbed
> where the owls sit humbly
> and the horses with lowered heads
> sleep in the star lit paddocks. (ll. 9-15)

A bittersweet aftertaste lingers with the reader. The fragrance "floats high clearing the creekbed | where the owls sit humbly." The cyclical reality of burning involves the transubstantiation of the tree, which returns to its generative habitat in a dematerialized form (see Chapter 2). "Sandalwood" commingles a domestic family memory with a poignant rite—the holding of vigil over the last stick of a rapidly diminishing plant species. Although sandalwood is not currently extinct, its original range has been severely reduced by land clearing. The tree's fragrance drifts like a *pneuma* from the playroom; smell transcends the confines of the domestic and social conditions that have caused its scarcity in the bush. On the whole, Hewett's poem hinges on a shift between the domestic habitat in which sandalwood is consumed or exploited and the habitat in which the tree is born, lives, and dies regeneratively, before the cutters "systematically" (l. 4) cleared it away as an export commodity—as a plant material.

"Sacrificing the Leaves" by Glen Phillips (1988, 8) also considers South-West flora from a domestic perspective, as Hewett does, but from "the green lawn" (l. 3) rather than the inner quarters of a playroom. In their poetry, both Hewett and Phillips wend between a number of themes, including the habitats of indigenous plants, their cultural constructions, their social histories, and the impositions of technology on regional ecologies. Phillips' poem opens with a familiar observation of the contrariety of Western Australian trees in comparison to their counterparts in the northern hemisphere:

> They say here the world's upside-down,
> And in summer it is true I find
> All the green lawn covered in the morning
> With this close pattern of what seem autumn leaves. (ll. 1-4)

Australian nature as "upside-down" is a trope dating back to early naturalists like James Edward Smith who published the first book on the flora of the new colonies (Moyal 1986, 20). "Sacrificing the Leaves" shifts from the contrariness of withered leaves on a green summer lawn to the contemplation of the ecological purposefulness—the autopoiesis—at work behind "this close pattern of what seem autumn leaves:"

> The eucalypts, wiser than the trees of the old world,
> Ancient in sacrificing to the sun what is its due,
> This way will find new strength to put out afresh

Tawny young leaf-sprays when the first autumn rains come. (ll. 5-8)

As a seasonal ritual, eucalypts defoliate in the summer in order to re-leaf during the first autumn rains. In Australia, the order of botanical nature is reversed, but not without reason; the act of sacrifice is an ecological one. "Sacrificing the Leaves" reflects the poet's ecological observations and knowledge of plant species, producing a poetics of indigenous South-West trees. As such, the poem can be considered habitat writing that evokes the continuum between the domestic contexts of lawns and the ecological settings of bushlands. Moreover, as with Les Murray's "The Gum Forest," the predominant tone of Phillips' poem is reverence—"the sapient eucalypt" (l. 13). Moreover, the tree inspires didactic metaphors for interpersonal relations:

And I think of lovers making their own source of light
And how they worship in its warming rays
And how, if they never learn the arts of sacrifice,
Love does not last a season; falls in the autumn days. (ll. 21-24)

Finally, reflecting aspects of ecology, sense, and science, the poem connects bodily movements to the arboreal phenomenon being observed: "I walk in the morning sun under the great trees | And my shoes thrust aside the fallen leaves" (ll. 18-19). In synchronizing physical movements and emotional cadences to ecological patterns, "Sacrificing" provides an example of a habitat poem with strong domestic themes.

Like Hewett and Phillips, poet Tracy Ryan (2002) was born in Western Australia. The flora and fauna of the Wheatbelt figures prominently in her writing. "Mallee Root" (2002, 16) mediates ecological consciousness of flora and barbed memories of social alienation. *Mallee* is defined as "a growth habit in which several to many woody stems arise separately from a lignotuber; usually applied to certain low-growing species of *Eucalyptus*" (Paczkowska and Chapman 2000, 579). The exhumed mallee root symbolizes the destruction of the Wheatbelt habitat by clearing and pesticides. The split between the country and the city forms the underlying breaking points of the poem. The misperception of indigenous vegetation reflects upon the livelihoods of Aboriginal people and diasporic inhabitants.

"Mallee Root" is more than a pure ecological meditation on mallee habitat. The poem opens with the paradoxical nature of an unearthed root mass, so unusual it defies the very notion of woodiness:

Not what we understood
as wood, this warped
and twisted thing
that had lain hidden. (ll. 1-4)

The root is a warped and twisted object, but not a piece of wood as one would know it. By the conclusion, the mass is described as "slow burning" (l. 23). The poet inserts pop cultural references to board games and television shows—*Lost in Space*, *Squatter*, and *Monopoly*—to create a sense of uneasiness in one's place.

These references are juxtaposed joltingly to the mallee formation itself "unravelling, | Medusa self-petrified" (ll. 21-22) to underscore the perception of the land as a plunderable resource—something as transient as popular fads. The mallee root and its habitat constitute what Elliott (1967, xi) refers to as a "metaphorical landscape," but the mallee is also a cultural milieu. As a grotesque Medusa head, the mallee root symbolizes the undervalued and misconstrued original flora of the Wheatbelt. In fact, after the flora had been cleared, the land could no longer hold back the vast underground salinity now plaguing the surface soil of the region (Beresford et al. 2001). The only value of the land, it seems, is as an object for drawing classes. The head is a thing of speculation:

> dry truffle, under a surface
> we knew only from
> *Lost in Space*, the jaundiced
> wastes stippled with sheep-
> skulls plundered by our city teachers
> for our 'contour drawing'. (ll. 5-10)

Human appreciation of Wheatbelt ecology is reduced to images of a science fiction television series. Ryan uses the mallee root as a potent metaphor of critique against the battery of misinterpretation that results from reading the surface appearances of a place rather than being with the ecological processes at work: "Smug lump presented like | a fact they had on us" (ll. 15-16). Like Hewett's "Sandalwood," Ryan's poem connects a childhood memory to the indigenous flora of the poet's home sphere. Recollections of "contour drawing" classes (l. 10) and "The Farm" (l. 14) provide critical anchors bridging ecological and cultural consciousnesses. The misunderstandings and stereotyping are directed not only at the mallee root but at the people who inhabit the mallee terrain—the dwellers in mallee place, of past and present.

Conclusion: Habitat Poetics

As these excerpts reveal, South-West habitat poets engage with ecology, sense, and science. A thread unifying the poetry of Lansdown, Choate, and Kinsella is conscious critical awareness of the scientific conventions surrounding plants. Indeed, integration between poetry and science is one of the trademarks of habitat poetry, including that of Hewett, Phillips, and Ryan. However, the sensory language used in some of their poems vis-á-vis plants at the same time deconstructs the exclusivity of scientific knowledge by embodying the poet as mediator in the environmental network of the plant. I have suggested that *habitat* is more than language or discourse; it is a way of interacting corporeally with the world—and with plants in this instance—through the many senses, just as landscape is a way of regarding the earth visually, as "a composition of that world" (Cosgrove 1998, 13) or as an act of "descriptive picture-making" (Elliott 1967, xi). In "Borrow Pit," Kinsella (2005, 142) concludes with:

> I borrow words
> from before I could speak, the tones of wandoo and mallee

>   intricacies of roots, and palettes of gravel
>   that stare us in the face, trunks horizontal, parallel
>   to the rippling undersurface, those winning ways. (ll. 87-91)

Attending to "the rippling undersurface" and the "intricacies of roots"—in other words, the complexities of the unseen—habitat poetry infuses language with ecology, sense experience, and the nuances of botanical knowledge. Unlike poetic language that uses highly visual tropes, the six poets featured in this chapter write across ecology, botany, process, place, memory, and sensation.

The selected works of these South-West poets help to underscore habitat poetry as a form of nature writing concerned with associations between a plant and other plants, animals, rocks, elements, and people. Yet it should be noted that not all habitat poetry is alike. For instance, Choate's "Poverty Bush" is a trenchant commentary on the idiosyncrasies of naming, but few of his poems deal with corporeal sensations. Lansdown's "A Few Weeks Later," in contrast, gives a playful critique of scientific knowledge through the poet's embodied experience. Invariably, the botanical poetry featured in this chapter is more multi-layered than the category of landscape poetry can allow. Hence, there is a need for *habitat* as a ecopoetic framework. In a biodiverse place like the South-West of WA, poetry has the capacity to instill appreciation for the botanical world by shaping the ways in which people regard plants and by distilling ecology into a palatable and palpable form. Through a blend of sensory, scientific (that is, taxonomic and classificatory), and ecological language, this cross-section of habitat poetry celebrates the adaptive successes of the South-West's indigenous plants. Building on this discussion, the next chapter applies concepts of ecopoetics and embodiment to analyze mycological writings from Australia and the United States.

Chapter 7: Four Poems on Mushrooms: A Poetic Mycology of the Senses

Introduction

Only 7 percent of the planet's fungi have been classified by science, compared to 90 percent of the planet's plants. If fungi are "truly the forgotten kingdom," then fungi poetry might truly be the forgotten ecopoetry. In my an analysis of four poems simply (and generically) called "Mushrooms," I bring together concepts of fungi, human sensoriality, multispecies theory, and ecopoetry. In the context of a poetic mycology, I draw from J. Scott Bryson's three features of ecopoetry. Bryson argues that ecopoetry, if it can be generalized, reflects an ecocentric perspective; an imperative towards humility; and a distrust of hyperrationality. The Australian poet Caroline Caddy's "Mushrooms," the American poet Mary Oliver's "Mushrooms," the American poet Sylvia Plath's "Mushrooms," and the American poet Emily Dickinson's "Mushroom" evidence, in different ways, a "poetic mycology of the senses." Oliver's poem is about human discretion with potentially hazardous mushrooms. However, her poem also points to intimacy as a way of navigating the mycotal world through the senses. Caddy's perspective on mushrooms is compellingly embodied. Plath, in comparison, writes from the mushrooms' point-of-view. Dickinson moralizes the mushroom. Using these four poems as catalysts, I develop the notion of a poetic mycology of the senses through multispecies theory and its relevance to ecopoetry. Haraway's idea of "companion species" and Tsing's "arts of inclusion" call for entangled sensorialities with nonhumans of "significant otherness"—which I believe is exemplified by mushrooms. As the three poems imply, the label *mushroom* is an ontological gestalt that belies the sensory complexities of fungi. It is through smell, taste, and touch that the radical otherness—the bewildering diversity and vexing ecologies—of fungi are made intimate and immediate.

Mycotal Otherness in Science and Ecocriticism

If fungi comprise "the forgotten kingdom," then poetry that takes fungi and the discipline of mycology as its subject matter could be—by association—the forgotten ecopoetry (or perhaps "mycopoetry"). As the third "f" in contemporary biodiversity conservation, languishing behind fauna and flora (Pouliot and May 2010, Walker 1996), fungi occupy a comparably liminal and, possibly, marginal position in literary history and ecocritical studies (for emerging work in this area, see Money 2011, Ryan 2012d, Tsing 2011). In particular, fungi straddle a largely unnavigated terrain between the relatively young field of "human-animal studies" (Shapiro and DeMello 2010) and its literary counterpart "zoocriticism" (Huggan and Tiffin 2010, 133-202) and the emergent "critical plant studies" (Marder 2013) and its budding complement "vegetal ecocriticism" (Adamson and Sandilands 2013). As a consequence, even among ecocrits, fungi have been grouped into the latter category, mirroring a tendency in the history of the biological sciences to aggregate fungi and plants. For example, Adamson and Sandilands (2013) refer to "floral, botanical, arboreal, fungal, and other *vegetal* discourses [emphasis added]." Yet, as neither plant nor animal—that is, existentially in-between the other two "f's—fungi lack the powers of photosynthesis synonymous with green plants, and also proliferate through radically different mechanisms (Griffin 1994). I, therefore, suggest that the ecocritical reading of mycotal poetry should be performed in the context of the unique otherness of these organisms.

In light (or is it the dankness?) of this, I ask in this chapter: What are the diverse ways in which human beings perceive fungi? What are the common figures of speech used to express the particular mycotal mode(s) of being? And more precisely: When does poetry shift away from hackneyed mushroom metaphors towards a curiosity for the complex lives and cultural meanings of fungi, as well as their irreplaceable ecological and social roles? In responding, I begin with a broad premise: by virtue of the ecologically and ontologically articulated modes fungi inhabit, to write of them is to write in a different way than of animals and plants. Indeed, despite the lack of parity between the kingdoms, fungi and particularly fleshy macrofungi or mushrooms, appear regularly in North American, European, Australian, and South African poetry (Ryan 2012d, Millar 2002, Roehl and Chadwick 2010). However, in asking "when" such a shift occurs in the history of poetry towards a view of fungi as relational and complex beings, I do not mean to undertake a historiographical analysis of mycotal writing throughout these traditions; such a project would be entirely out of my present scope. Instead, I do wish to know what the gestalt label *mushroom* signifies for these four somewhat disparate poets; the ways in which Dickinson, Plath, Oliver, and Caddy comparatively represent mushrooms in language; and some plausible reasons for the differing qualities they attribute to fungi and the manner in which they do so.

In my analysis of four poems generically (and plurally) titled "mushrooms" (or in Dickinson's case, the singular form, "mushroom"), I will take note of the recurrence of mycotal tropes, while considering the implications of such ways of regarding fungi for broader cultural perceptions of the kingdom. Indeed, as we will see, in much poetry about fungi, mushrooms are linked symbolically with danger, decay, and death, as well as stealth, sin, and the supernatural. These attributes operate as stock tropes, reinforcing certain largely negative preconceptions about this much misunderstood, disregarded, and "forgotten" group of beings. Accordingly, in my readings, I will be on the lookout for positive representations of human-fungus entanglements—sensory intimacies if you like—involving the apprehension of kingdom Fungi as a community of beings through the speaker's direct embodied experience. In these rare instances, the eating, tasting, smelling, and touching of the delectable fruiting bodies of mushrooms leads to what French philosopher Michel Serres calls a situation of "mingled bodies" (2008) in which physical and intellectual distance between humans and fungi dissipates—along with human mistrust—if only fleetingly. Building on Serres and others, as part of this admittedly brief comparison offered in this chapter, I will draw upon ecopoetic and multispecies theory in order to conceptualize the implications of these poems more generally for kingdom Fungi.

Theorizing a Poetic Mycology of the Senses

Through these four poems on mushrooms, the notion of a "poetic mycology of the senses" will be forwarded and, to some extent, developed as a critical lens for reading mycotally focused environmental writing. In particular, Scott Bryson's elaboration of the three features of ecopoetry, in conjunction with Scott Knickerbocker's productive notion of "sensuous poesis," will be used to explore the ecopoetic foundation for a poetic mycology. Bryson argues that ecopoetry bears three distinguishing attributes. To begin with, ecopoetry reflects "an ecocentric perspective that recognizes the interdependent nature of the world" (Bryson 2002, 5-6) or, in Timothy Morton's terms, reflects ecology as "thinking how all beings are interconnected, in as deep a way as possible" (Morton 2010, 255). Secondly, as Bryson goes on to explain, ecopoetry expresses "an imperative toward humility in relationships with both human and nonhuman nature" (2002, 6). And thirdly, ecopoetry reveals an abiding suspicion of "hyperrationality and its resultant overreliance on technology" (2002, 7). In Bryson's framework, hyperrationality refers to the preponderance of deductive logic and analytical reasoning to the exclusion of other modes of knowledge-making, including sensory embodiment in a place, intuition, empathy, and the interrelationships between species. The operative term in Bryson's analysis is *overreliance*. In fact, technological instruments, such as electron microscopes, can facilitate sensuous human encounters with fungi and their physiologies that would otherwise be impossible to the naked eye. However, *sensuous poesis* refers to a condition of multisensoriality, combining the powers of vision with the nuances of tasting, smelling, touching, and hearing.

As the underlying foundation to Bryson's three attributes, human sensory embodiment in the material domain helps to make possible an ecocentric attitude, commitment to humility, and skepticism in the face of hyperrationality. Citing the aurality of contemporary American poetry, Knickerbocker augments Bryson's three-fold position through the term *sensuous poesis* as "the process of rematerializing language specifically as a response to nonhuman nature" (Knickerbocker 2012, 2). According to Bryson, sensuous poesis inverts the mirroring of the world in language (as pure representation) and rather inflects the immanent sensory potential of poetry to "enact, rather than merely represent, the immediate, embodied experience of nonhuman nature" (Knickerbocker 2012, 17). My sensory analysis of mycopoetry adds another dimension to Knickerbocker's notion of sensuous poesis as an outcome of language. Moreover, the material exchange between mingled bodies (to invoke Serres) involves human-mushroom interpenetration that disrupts aesthetic or linguistic distance and stanches negative moral attachments directed at mushrooms. In short, the interrogation of language is essential to understanding mushrooms and redefining human-fungus relationships.

In addition to the ecopoetic theory of Bryson and Knickerbocker, multispecies theory proffers another lens for comprehending what these poems reveal about human-fungus entanglements. Multispecies theory encompasses a body of writings by posthumanist scholars that sets out to decenter human subjectivity and to value the multiple subjectivities—of animals, plants, insects, and mushrooms—of the ecocultural (i.e. environment + culture) world (Haraway 2008, Morton 2010, Wolfe 2010). Entanglement is an integral notion within multispecies theory—one which implies a degree of sustained material reciprocity between the mingled organisms and their lifeworlds. The term *lifeworld* derives from the philosophical writings of Edmund Husserl and describes a world experienced in common by all living beings: plants, animals, fungi, and humans alike. For a lifeworld to exist and to lead to knowledge, the multiple senses must be engaged in a sustained way with our "surroundings." One is necessarily entangled with one's lifeworld; one *is* one's lifeworld. Barad stresses that "entanglements are not a name for the interconnectedness of all being as one, but rather specific material relations of the ongoing differentiating of the world. Entanglements are relations of obligation—being bound to the other—enfolded traces of othering" (Barad 2010, 265).

Moreover, Anna Tsing's "arts of inclusion" (2011) and Donna Haraway's "companion species" offer conceptual models for articulating the multiple entanglements between humans and nonhumans of "significant otherness." Haraway highlights the co-constitutive sensory dimensions of companion species or "the many tones of regard/respect/seeing each other/looking back at/meeting/optic-haptic encounter. Species and respect are in optic/haptic/affective/cognitive touch" (Haraway 2008, 164). Foregrounding the etymological linkage between *species* and *respect*, Haraway's term allows for the consideration of "which categories are in play and shaping one another in flesh and logic in constitutive encounterings" (2008, 164). More so than flora and fauna (especially charismatic furry animals and venerable old trees), non-human otherness is exemplified in the third "f," kingdom Fungi: the slimy, stealthy, secretive, subversive, and sinful. The task of rethinking fungi begins with critically regarding the categories employed to constitute them and the language used to do so.

In tandem with ecopoetic and multispecies theory, these four poems act as catalysts for a poetic mycology. Emily Dickinson's "Mushroom" (1874), Sylvia Plath's "Mushrooms" (1960), Mary Oliver's "Mushrooms" (1983), and Australian poet Caroline Caddy's "Mushrooms" (1989) exhibit different aspects of the notion—three of which I will highlight and develop in this chapter. The first aspect refers to the representation of fungi in language through the commonplace tropes—such as physical decay and moral decrepitude—that are (often wrongly) applied to express mushroom beingness and our *being with* mushrooms. The second articulates the degree to which the ecology of fungi factors into the ecopoem, demonstrating "the interdependent nature of the world," in Bryson's terms or entanglement as "specific material relations", in Barad's. The third and most prominent aspect I will touch on relates to how an ecopoem materializes human sensory embodiment through the interplay of the autocentric (smell, touch, and taste) and allocentric senses (sight and hearing) in language (Porteous 1996, 31). Dickinson, Plath, Oliver and Caddy's poems exhibit differing intensities of bodily interaction with fungi. Their poems collectively represent a continuum of human-fungus interaction—from the distanced and demonized mushroom of Dickinson to the sensuous edible species of Caddy that facilitate the reconciliation of a troubled mother-daughter relationship.

The Elf of Plants: Emily Dickinson's "Mushroom"

In 1830, Emily Dickinson was born in Amherst, Massachusetts, USA, where she later died in 1886 at age fifty-five after a notably reclusive life. Ecocritics have observed the sensitivity to the environment that is integral to her poetic oeuvre, as well as the revisionist qualities of her nature poetry in contrast to the largely masculinist Romantic and Transcendentalist visions of nature preceding and contemporaneous with her (Felstiner 2009, Stein 1997, Chapter 1). As Stein comments, "her nature poetry addresses and undermines the prevailing masculinist assumptions about women and nature espoused by the Romantic and Transcendentalist writers and by Puritan theologians of her day" (1997, 25). Dickinson's "Mushroom" (1874) evidences her astute perception of the natural world, evoking fungi as supernatural beings in its opening stanzas. Yet its conspicuous puritanical themes overshadow the environmental relationships of the mushroom (of course it is unlikely that Dickinson would have recognized the word *ecology* in the first place) and, unfortunately, occlude any form of sensory involvement with the organism. In the poem, the mushroom is a dangerous Judas-faced entity lacking both moral consideration and capacity. Written by a middle-aged Dickinson, the version quoted here retains the idiosyncratic capitalizations of the author's original and includes revealing word choices that are altered in subsequent published versions (Franklin 1999, 520). As mycologist Nicholas Money argues, the majority of mycotal poetry—Dickinson's being no exception—leverages connotations of danger, death, and decay in reference to fungi, particularly linking them to witchcraft and the divine (Money 2011, 136). Money specifically observes Dickinson's use of the image of Judas in the final stanza, reiterating everyday negative symbolic associations between mushrooms, morality, and religious institutions (2011, 134). Hence, the poem's pejorative tone could directly reflect the Puritanical mood of nineteenth-century New England or, alternately, could be interpreted as a thinly veiled acerbic commentary by Dickinson on the moralization of nature by American religious institutions of her era.

Associations between the supernatural world and mushrooms appear in the opening verses: "The Mushroom is the Elf of Plants - | At Evening, it is not | At Morning, in a Truffled Hut | It stop opon [sic] a Spot | As if it tarried always" (ll. 1-5). An elf is an archetypal otherworldly being, used rather unsurprisingly by Dickinson, but it is also an ambivalent figure, a shape shifter, a changeling, transmogrifying through a timescale dramatically different to human temporality: "At Evening, it is not." Whereas the first stanza concerns the mushroom's supernatural qualities, the second turns to the brevity of the organism's lifespan, the spontaneity and erratic nature of its growth habits, and the unhuman biorhythm it manifests through its cryptic and furtive movements: "And yet it's [*sic*] whole Career | Is shorter than a Snake's Delay - | And fleeter than a Tare -" (ll. 6-8). Following Dickinson's assessment of the mushroom's occultism, manifested by its inhumanly mannerisms, the poem shifts in the third and fourth stanzas to a multitude of associations. These primarily serve to connect the mushroom to sorcery, deception, secrecy, and evanescence, relegating it to a "surreptitious Scion" or, in other words, a fungus on the sly, an inferior plant or, worse yet, a biological poser for the vegetal.

Metaphors such as "Vegetation's Juggler" (l. 9), on one hand, characterize the mushroom as akin to a circus act performer—a participant in something not legitimate, not real, definitely not categorizable. On the other, such a phrase suggests that fungi are shapeshifters existing outside of the visible—decomposing, connecting, transforming or, in other words, orchestrating the perceivable and familiar ecological forms of shrubs, trees, animals, and soil. Read negatively, however, these tropes also conjure the trickster figure, the mesmerizer, and the slight-of-hand charlatan. Moreover, the phrase "Germ of Alibi" (l. 10) implies the microorganism theory of disease of the late nineteenth century, perhaps known by Dickinson at the time of writing, which would have implicated fungi with a multitude of afflictions and adverse states of health. Morally, the phrase connotes the evasion of responsibility for the committing of evil or criminal acts. The final stanza is even more condemning than the first four: "Had Nature any supple Face | Or could she one contemn - | Had Nature an Apostate - | That Mushroom - it is Him!" (ll. 17-20). An "Apostate" is someone who abandons his or her religion, who defects from institutions of worship, who converts for the *worst*. The term is denoted in a later version of the poem as "Iscariot" or Judas Iscariot, who betrayed Jesus with a notorious kiss and thereby became the archetypally deceptive and faithless persona in Christian doctrine.

While demonstrating canny observation of the natural world around her, Dickinson's poem unjustifiably excoriates "The Mushroom," projecting toward it a battery of unflattering associations. The mycotal tropes used by Dickinson personify the mushroom as incontrovertibly deceptive and evil—as a being culturally misunderstood for its perceived secrecy, stealth, and sorcery. Additionally and perhaps most condemningly, the mushroom is not part of nature: "Had Nature any supple Face | Or could she one contemn" (ll. 17-18). The mushroom is nature possessed, in Dickinson's terms, much as a cancer is a plague on the body by the body. The tone is distanced—the speaker's relationship to the mushroom and also the reader's subsequent regard for the mushroom—and its relentless barrage of moral associations obscures the ecological dynamism of fungi, or in Morton's terms, ecology not as science per se, but rather as the ongoing consideration of "how all beings are interconnected, in as deep a way as possible." There is merely one multispecies allusion—"I feel as if the Grass was pleased | To have it intermit" (ll. 13-14)—hinting at an awareness of the interactions between the mushroom and its living environment. However, Dickinson's use of "surreptitious Scion" (in plant propagation, a living part used for grafting) aggregates fungi and plants, a conflation that obscures the unique *umwelt* of mushrooms and relegates them to imperfect plants, perpetuating a long-standing bias that degrades fungi as failed flora. An alternate, ecological reading of "surreptitious Scion" would acknowledge the mycorrhizal associations between fungi and plants in which fungi symbiotically extend the reach of the grasses.

Dickinson's poem is representative of the gamut of symbolic meanings attributed to mushrooms, especially those of the poem's historical moment. It falls short of offering a poetic mycology of the senses in the three interlinked dimensions I propose: linguistic, ecological, and sensorial. In sum, while there are elements of sensuous poesis, there is little indication of human-fungus interactions through intimacy and entanglement of any sort. In the final analysis, Dickinson's "mushroom" is generic (although it assumes problematic "faces" throughout the poem). It is indistinguishable from the masses, "fleeter than a Tare" (l. 8) and representative of collusion—a defeated object associated with the figure of Judas as a focus of moralization and proselytizing. Despite the singular form of the noun, the mushroom is neither an individual (in the sense that an animal is an individual) nor a collective (in the sense that an individual fruiting body is part of a vast underground network or mycelium). What is missing in Dickinson's rendering of the mycotal is a sensory, ecological, and imaginative interest in mushrooms for their own sake (apart from their religious and supernatural faces), one which closes the human-fungus yawn wrenched open by continuous misunderstanding and inappropriate moral attribution. Admittedly, my diachronic reading, beginning with Dickinson's poem, is not meant to show a progression of "bad" to "good" mycotal poetry but rather to demonstrate the shedding of certain symbolic attachments and the subsequent re-envisioning of fungi for what they *are* and for what they *can be*.

Earless and Eyeless: Sylvia Plath's "Mushrooms"

Sylvia Plath was born in 1932 in Boston, Massachusetts, USA, and died in 1963 at the age of thirty-one. She is typically interpreted by literary critics as an intensely interior and distraught confessional poet. For example, Holbrook (1976, 269) offers an existential reading of "Mushrooms" as a poem "which couldn't […] have been written without the torment of experiencing a life without feeling alive." However, Plath's poetry has been considered for its environmental consciousness in Tracy Brain's full-length study of the poet's worldliness—specifically in response to the pesticide toxicity brought to widespread attention by Rachel Carson's seminal book *Silent Spring* (Brain 2001). Normally examined for its painful introversion, Plath's verse seen in a different light expresses the relational, multispecies, and corporeal ethos at the core of posthumanism and multispecies theory. Similarly, Knickerbocker (2012, 126) argues that Plath "expresses the ecological idea that death is often linked to alienation from one's environment and fellow creatures, whereas life requires interaction with one's environment and other beings." Nevertheless, Knickerbocker's statement suggests a binary between death and life that does not hold well in the context of fungi. As saprophytes, detritivores, and decomposers, fungi are intrinsically linked to death and even thrive under conditions of decay. Fungi allow us to realize that both life and death require "interaction with one's environment and other beings" and that embodiment in the world is a condition of life *within* death and death *within* life. Rather than one-dimensionally death-obsessive, Plath's poetry reflects this complexity as a "desire for sensuous embodiment"

(Knickerbocker 2012, 127) through direct experience and acute awareness of nature. Her poetry manifests the notion of *sensuous poesis*, as Knickerbocker (2012, 134) goes on to say, in that Plath's "intense imaginative capacities were not simply a matter of artistic intention but were also a nearly bodily compulsion."

In "Mushrooms" (1960), she grants an imaginative perspective to fungi, personifying them, giving them intentionality, and allowing them to speak for themselves as a collective. As Knickerbocker (2012, 135) also cogently observes of the interwoven imaginative and material dimensions of the poem, "Plath's use of first-person plural is not merely a poetic flight of fancy; it expresses an ecological verity"—the underground mycelium of mushrooms that constitutes a single organism composed of many, a creature of communalism. Unlike Dickinson's demonized archetypal mushroom with Judas facelessness, Plath's mushrooms take the form of an interconnected being not existing in isolation but rather, to borrow Jean-Luc Nancy's (2000) term, as "being singular plural." Interpreted through the lens of environmental embodiment and Knickerbocker's sensuous poesis, Plath's "Mushrooms" and other poems from her oeuvre narrate a process of human absorption into ecology whereby one's body is subsumed within the materiality of the natural world. Edward Butscher (1976, 244) comments appositely that the poem shoves "her consciousness directly into the eye of nature itself."

Plath's poem begins with the human perceptions of mushrooms that are commonplace to other mycotal writings, including quietness, stealth, and sudden appearance from nowhere, or so it seems: "Overnight, very | Whitely, discreetly, | Very quietly" (Plath 1967, 34-35, ll. 1-3). Yet, the fungal form adumbrated by Plath, despite its otherness, is conspicuously human: "Our toes, ours noses | Take hold on the loam, | Acquire the air" (ll. 4-6). These mushrooms, courtesy of her mycological imagination, have recognizable appendages as well as an animal-like capacity for respiration. Their anthropomorphic attributes are reiterated in "Soft fists insist on | Heaving the needles, | The leafy bedding | Even the paving" (ll. 10-13). As such, the mushrooms' corporeality opens up the possibility of bodily empathy between kingdom fungi and human beings. That "Nobody sees us, | Stops us, betrays us" (ll. 7-8) invokes again the slyness and abruptness of their arrival—the particular timescale of their movements that contrasts starkly to mammalian motion. Despite an incomprehensible temporal rhythm, the dynamism of the mushrooms is celebrated in the poem as they physically tousle the leaf litter and subvert the pavement, pushing upward with their "Soft fists" (l. 10) as "Our hammers, our rams" (l. 14).

Navigating without the allocentric senses of hearing and sight—indeed they are "Perfectly voiceless" (l. 16)—the mushrooms' dynamism is distinctively tactile as they heave, shoulder, nudge, and shove their way upward. They burst forth, pry open doors, widen crannies, shoulder through holes, and heave the needles—active phrases that convey their enervated activities. On the whole, Plath's poem expresses convincingly the plurality of the mushrooms—"So many of us! So many of us!" (ll. 23-24)—that "In spite of ourselves. | Our kind multiplies" (ll. 29-30). From the first-person plural perspective (indicated by "our," "us," and "we"), there is a prevailing sense of mushrooms constituting an ecological community—mushrooms as a singular mushroom in dynamic relation to its plurality, as the collective voice of many heaving upward bodily together in overwhelming profusion. Furthermore, their dynamism is also the juxtaposition of "hard" attributes and forms ("We are shelves, we are | Tables...") (ll. 25-26) and soft, malleable qualities (we are meek, | We are edible) (ll. 26-27), the latter importantly signifying the potential for humans to eat these kinds without consequence.

The enigmatic final tercet recalls Dickinson's biblical reference in "Had Nature an Apostate - | That Mushroom - it is Him!". Plath concludes: "We shall by morning | Inherit the earth. | Our foot's in the door" (ll. 31-33). However, the conspicuous allusion to the Book of Matthew, "Blessed are the meek: for they shall inherit the earth," operates in a manner that affirms mycotal being and is ultimately undergirded by the tenacious embodied presence of the mushrooms themselves, animatedly prying open yet another human-constructed space: "Our foot's in the door" (l. 33). If ever mushrooms were relegated to a forgotten kingdom, inhabiting a haunted position within the Western cultural imagination, their meekness, silence, and discretion (in the opening tercet) culminate over the poem's timeframe in a dynamic force (in the final two tercets) that overcomes their merely being overlooked or inadvertently stepped on, as the final line intimates. Of course, the poem's emphasis on multiplicity could have gone the direction of pathogenic excess, but a feeling of awe and reverence lingers. Although the species identity is not revealed, we accept that they are mushrooms, maybe the common edible field variety. It could be that "mushrooms," for Plath, is a composite signifier standing in for different kinds of fungi, both edible and poisonous, subterranean "fists" and tree-borne "shelves" and "tables."

The generalizability of the term, therefore, works positively in the poem, allowing the diversity of kingdom fungi to be voiced imaginatively (and cacophonously) in chorus. The generic appellation "mushrooms," as a gestalt category generated at the margin of human awareness, on "crumbs of shadow" (l. 20), becomes, by the poem's end, a vociferous and inescapable concerto. In contrast to Dickinson's mushroom, Plath distinguishes her perspective on the mycotal world by celebrating mushrooms for their tenacious qualities or, in Money's (2011, 135) terms, "the steady, inconspicuous development of the fungus before its glorious fruiting as a metaphor for patience and self-possession, assertiveness, and activism" and, I add, intentionality. Quite convincingly, we see in the poem the overturning of the stock pejorative associations between mushrooms, social parasitism, and rapacious growth toward a poetic mycology of the senses from the mushrooms' point-of-view. In its equating of "a certain segment of animal or vegetable [or mycotal] life with human existence" (Butscher 1976, 248), "Mushrooms" forwards the multispecies momentum toward decentered human subjectivity rather than pernicious human solipsism.

Flocks of Glitterers: Mary Oliver's "Mushrooms"

Born in 1935 in rural Ohio, USA, Mary Oliver settled in Provincetown, Massachusetts, later in life where much of her poetry is set. Reflecting her immense curiosity for the world, many of Oliver's essays and poems address themes of ecological interdependence, intimacy with nonhumans, and the immediacy of direct experience. In his reading of Oliver's "pragmatic mysticism" and the relational attributes of her work, Laird Christensen (2002, 137) observes that "traditional distinctions between mortality and immortality quickly break down in Oliver's poems as the material elements of each being are transformed into the elements of other bodies." In terms of human-nature entanglement, Oliver's poetry exhibits a "continual reintegration of [the] individual into the whole [that] denies any abiding sense of discrete identity" (Christensen 2002, 137) and thereby decenters human subjectivity by placing the activities of people within a material, ecological community marked by cycles of growth and decay, life and death. Oliver's poem opens with the ecologically rationalized emergence of mushrooms—demystifying their sudden arrival, attributed as we found in Dickinson's poem, to the workings of the supernatural rather than habitat processes: "Rain, and then | the cool pursed | lips of the wind | draw them | out of the ground" (Oliver 1992, 144-145, ll. 1-5). Instead, in the poem, the convergence of elements—moisture, temperature, wind, and earth—galvanizes the appearance of mushrooms, nonetheless death-evoking for Oliver. As in many of Oliver's ecopoems, the materiality of the mushrooms in their milieux becomes an ecological force with considerable physical momentum and apparent dynamism: "red and yellow skulls | pummeling upward | through leaves, | through grasses, | through sand…" (ll. 6-10).

Like Dickinson and Plath before her, Oliver represents the habitus of mycotal being-in-the-world as closely in synch with time and sound: "astonishing | in their suddenness, | their quietude, | their wetness, they appear | on fall mornings…" (ll. 10-14). Yet, unlike Dickinson and Plath's poems, while some mushrooms are "packed with poison" (l. 17), others are "billowing | chunkily, and delicious" (l. 18-19). In Oliver's work, we are presented with a more equanimous picture of the fungi kingdom, as both death-dealer and life-giver. Through the physicality of walking amongst the flocks, the human capacity for discernment (and hence self-preservation) is fostered through close sensory interaction with mushrooms: "those who know | walk out to gather, choosing | the benign from flocks | of glitterers, sorcerers, | russulas, | panther caps, | shark-white death angels" (ll. 20-26). As such, Oliver's mushrooms are beyond the categories of moralization (of attributing goodness or evil to them) and, instead, exist as corporeal beings, whether edible or poisonous or in-between; indeed, to skirt death in the field, one must become one who knows of their physical properties. As a linguistic tactic employed by other nature writers on fungi, the likening of mushrooms to supernatural figures—"glitterers, sorcerers"—rather than weakening human-fungi entanglements in Oliver's poem, instead recalls Haraway's linkage between *species* and *respect* or "seeing each other/looking back at/meeting/optic-haptic encounter." In contrast to Dickinson's othering of the mushroom, which brews Judeo-Christian-based contempt by the poem's conclusion, Oliver's othering breeds respectful knowing, leading to secure delectation—the discerning between "sugar" (l. 28) and "paralysis" (l. 29).

On the whole, an uncanny mixing defines Oliver's "Mushrooms"—its movements polarized by the presence throughout of predictable preternatural tropes on the one hand (e.g. "glitters, sorcerers") and, on the other, a more sophisticated and specific lexicon (familiar to many field mycologists) with nuanced symbolic meanings and social resonances (e.g. "russulas, | panther caps"). In short, Oliver's enumeration of names, such as death angels, narrows the identities of these mushrooms and provides a basis for differentiating the virulent from the innocuous amidst the plurality. Despite a somewhat contradictory trajectory through kingdom fungi, the poem resounds a clarion message that familiarity and intimacy—intrinsic to Haraway's notion of companion species—grow through respectful human-fungi interactions in which the dangerous potential of some species is recognized, learned, and avoided. To state it differently, the intimate act of eating mushrooms—as experienced wild-crafters would know—necessitates the practical ability to tell poisonous species from delicacies. The acquisition of knowledge about fungi, although represented in cryptic and cultish terms in the poem, is, therefore, based more firmly in the experience of the everyday material domain: the ground, the earth, the fields of rain. In their poisonousness, the russulas, panther caps, and death angels themselves are not typecast as morally culpable agents—in fact they are "being perfect" (l. 33), unlike Dickinson's reprehensible Judas-faceless mushroom. Instead, they follow the truth of their ecological cadences, receding "under the shining | fields of rain" (ll. 35-36), leaving the staggering down of humans, poisoned by the deadly toxic few, to human agency (choice, discretion, intelligence) alone.

Everywhere They Touched Us: Caroline Caddy's "Mushrooms"

The only non-American (and non-New Englander at that) poet of the four featured in this chapter, Caroline Caddy was born in Western Australia (WA) in 1944, but lived as a child in the United States and Japan (Kinsella 2009, 401). Her most recent collection, *Esperance: New and Selected Poems* (2007), features a variety of ecologically conversant works, such as "Stirling Ranges" and "Karri Trees," about the South Coast region near Albany, WA. "Mushrooms" from her earlier collection *Beach Plastic* (Caddy 1989, 32-37) is a five-part poem centering on the troubled relationship between a mother and teenage daughter. As with Plath and Oliver's examples, Caddy's poem emphasizes embodied human-fungus interaction, most actualized through eating. There is an evocation of mushrooms as companion species through, to apply Tsing's term, an art of inclusion: the wild-crafting and preparation of edible species, involving the bringing of the mycotal other into a domestic setting where its symbolic and material dimensions become manifold, where it achieves ritualistic and spiritually cleansing status. Dispensing with the stock tropes we find in Dickinson that denigrate kingdom fungi, Caddy's perspective on mushrooms is compellingly corporeal and touchingly intimate: "We made soup | wiped their photocopies from plates [...] tables – | everywhere they touched us" (Sect. 3, ll. 36-38). Here, the notion of a poetic mycology of the senses reaches its apotheosis, in which palpable entanglement between the speaker and the still enigmatic but highly respected mushrooms occurs on multiple levels.

After evocative depictions of cooking in Section 2, in the poem's Section 3, the speaker commands her daughter out of the house to go mushroom gathering: "Navigating the seas of your boredom | I sent you out to look for mushrooms. | You returned feet wet skirt held | hem to waist" (Sect. 3, ll. 1-4). The daughter's voluminous fungal findings affect her physical and emotional balance positively: "you leaned back from your impossible burden | grinning your if you believe it it won't be so | and if you don't | it might not be either grin | I was sure you filled your skirt | with sticks and litter" (Sect. 3, ll. 10-15). Through the fecundity of the mushroom harvest and the season of fungi, the two discover, if only momentarily, a renewed empathy for one another as the wild mushrooms co-occupy the domestic space. What follows is perhaps the most compelling example of sensuous poesis specific to the mycotal in which language enacts the immanent physical sensations of nonhuman nature:

> I smelled them before you opened your skirt –
> not rank that often comes with size
>   but redolent
> our words came out like inspired praise.
> They were
> bowls for thick-lipped giants shepherd pies
> mosques and edible turbans.
> They had the feel of gruyere and some
> with strips of grass tied over them
>   were obscure Japanese packages. (Sect. 3, ll. 26-35)

Gradations of smell fall between "rank" and "redolent," particularizing the nuance of sensory experience. Rather than fearful distance, familiarity (literally in relation to the speaker's family) and human-fungus intimacy inflect the excerpt throughout. Caddy's haptic tropes signify interactions towards the attainment of human nourishment: "the feel of gruyere" (l. 33). The making of soup symbolizes the reconciliation between mother and daughter, as well as an embodied entanglement between human beings and mycotal companion species. The third section of "Mushrooms," the most fungally focused, expresses uninhibited sensory openness to fungi—"everywhere they touched us" (l. 38)—combined exactingly with the practical expertise of a wild-crafter. Caddy's mycological imagination and material poetics conspire to liberate fungi from an obsolescent language that constrains these organisms with insinuations of the supernatural, sin, treachery, and deceit. Individual mushrooms receive lucid and imaginative faces—"Russian domes", "photocopies from plates", "obscure Japanese packages"—that identify them within the plurality of their masses and their generic appellation: mushrooms.

Conclusion: The Poetry of the Forgotten Kingdom

As this brief foray through four mycopoems suggests, the signifier "mushrooms" is an ontological gestalt that can belie the sensory complexities and individual nuances of mushrooms and human-fungus interactions. It is through the autocentric senses of smell, taste, and touch, in conjunction with practical, field-based knowledge of the Kingdom, that their radical otherness, bewildering diversity, and vexing ecologies can be made intimate and immediate. The dynamism of the poems of Plath and Oliver reflects the unique habitus of mushrooms, whereas Dickinson's earlier attempt appears mired in its own symbolic detritus and perhaps that of its time. In Oliver and Caddy's works in particular, a poetic mycology of the senses emerges through the striking combination of ecological sensitivity and bodily invocation—both fostered through the intimate act of eating mushrooms. Indeed, Caddy's is an exploration of consuming fungi and its social and family implications. In her poem, particularly the third section, we find the full (and exemplary) expression of a poetic mycology of the senses—interlinking linguistic forms, fungal field ecology, and sensory experience—toward the creation of novel and surprising modes of language communicating human-fungus interactions in all their stickiness.

As a final note, the criteria of edibility is only one facet of human-fungus relationships. In this context, Haraway's etymological connection between *respect* and *species* raises the notion of "deferential regard" that has been associated with the former term since the 1540s (Harper 2013). Learning to avoid potentially lethal fungi and to harvest edible kinds is both a matter of respect (for Other and self) and the preservation of one's life. Such learning is an ongoing pursuit in which respect for the agency of these organisms is tempered with care for self; it necessitates an entanglement (not necessarily a deference but an ethic of caution), one that comes as a result of prolonged attention to fungi in the field, concerted study of taxonomic knowledge, and genuine regard for the successful survival mechanisms of these organisms. As Dickinson's poem implies through its use of the Judas metaphor, mushrooms—wrongly identified or imprudently trusted—can kill us or make us dreadfully sick. Yet, the dangerous properties of a few species should not negate the complexity and importance of the Kingdom as a whole. Indeed, emerging ecological knowledge of fungi reveal other horizons for a poetic mycology of the senses: for example, the multifaceted role fungi play in maintaining habitat processes, such as the transfer of ions between terrestrial and aquatic habitats (Gadd 2000). These ecological perspectives will continue to mark the evolution of sensuous poesis founded in notions of science, embodiment, entanglement, respect, and being with.

Chapter 8: Plants, Processes, Places: Sensory Intimacy and Poetic Enquiry

Introduction

> Intimacy obtains only where the intimate—world and thing—divides itself cleanly and remains separated. In the midst of the two, in the between of world and thing, in their *inter*, division prevails: a *difference*. (Heidegger 1971, 199)

In 2008, I began research on the aesthetics of South-West Australian flora, later published in the book *Green Sense* (Ryan 2012a). Having previously studied environmental philosophy and design, I was aware of the problematic relationship between aesthetics and landscapes (for example, see Berleant 2005; and also Chapter 5 of this book). From the outset, I felt that aesthetic theory and its practices focused on the distillation of spaces into objects, rather than the exploration of human sensory intimacy with lived places. Vistas, views, vantage points, and pleasing prospects reduce land to the picturesque, transforming the fields of contact in which sensations happen into surfaces (Cosgrove 1998). Yet, not only a physical array to be admired distantly, places comprise felt intersections—those brushes with plants and animals, those encounters with non-human life arising from physical closeness. Immanent experiences of place—engaging the senses of touch, taste, and smell in particular—engender attachment and affection through the emotions and the body, as I have argued throughout *Being With*.

The Praxis of Being With

According to the environmental philosopher Allen Carlson (2000, xvii) (whose "natural environmental model" was explored in Chapter 2), aesthetics can be defined as "the area of philosophy that concerns our appreciation of things as they affect our senses, and especially as they affect them in a pleasing way." In my ongoing research on aesthetic experience of plants, I have argued for a model of a multisensorial aesthetics through the exploration of human encounters with plants in their habitats, or "cultural botany" (Ryan 2012a, Chapter 1). Working within the highly interdisciplinary academic fields of cultural studies and ecocriticism, I have attempted to develop a poetics of embodied place, of contact "in the midst of the two, in the between of world and thing," in Heidegger's terms. My field-based approach responds to the tendency in popular culture to reduce the botanical world to a consortium of beautiful objects. The *dif-ference* brings me closer to plants through poetic engagements that linger in the "middle voice" (Carter 1996, 331) toward the experience of connectivity gained through the senses. I prefer to think that I grow and change along with the plants I research. Rather than detached viewing of landscapes, I have taken an interest in gestures of smell, contacts involving taste, and proximities of touch—those untidy sensory entanglements with the world, or what Serres (2008) refers to as our "mingled bodies." For example, soundscapes and smellscapes, though invisible, can bring about sensation, especially as the senses interact among themselves and with the wor(l)d. As the geographer Paul Rodaway (2002, 64) maintains, smell evades the kind of spatial organization at the center of visual aesthetics: "Smells infiltrate or linger, appear or fade, rather than take place or situate themselves as a composition." However, as far as I can surmise from Kant, the term *taste* itself has become indicative of visual desirability, as a metonym for aesthetic judgement. Aesthetics, as I know it, seems to freeze living things in images as a science of perception.

In taking a position polemically for sensory interestedness and bodily entwinements, I noticed that my academic prose began to suffer from the same stasis I was arguing to counteract. I needed poetic insertion, moments of movement and sense infusion dimpling the carapace of theory I was hiding under. *Entry points*. The poetry would exhibit *praxis*, as derived from the Marxist tradition "implying not just practice, but the ideological assumptions undergirding and/or deriving from practice" (Phillips 1999, 599). For me, praxis is reflexive; theory is continually shaped by practice, and, conversely, practice by theory. The interludes in *Green Sense* are spaces of in-betweenness—interstices in the text—that ground my theoretical ratiocinations. Literally meaning "between play," an interlude intervenes between the longer segments of a performance (an academic book, in this case). Unlike an intermission that disrupts continuity with a pause, an interlude preserves the flow of the work. As part of my praxis, poetic interludes affirm my position, not as a disembodied voice speculating, but an embodied presence participating in the sensuous plant life about which I produce knowledge.

Interlude I: Lesueur National Park, Jurien, Western Australia

A jaunt through the knee-high vegetation atop Mount Lesueur north of Perth, WA in 2009 was my first excursion into the rich heathlands known as the *kwongan* (Beard and Pate 1984, xvii-xxi). "Understanding Parrot Bush" narrates an induction (in Fremantle Poets 3, edited by Mitchell 2013). My body-based investigation of the flora of the South-West region of WA began with *Dryandra sessilis*, also known controversially (as of 2007) as *Banksia sessilis*. Through the shifts between discord and ease in the structure of the writing, the poem expresses acculturation to a place through a curiosity about its flora. As a newcomer to the South-West, the quality of harshness for which Australia's plants are known, strikes me as the obverse of the silken texture of the flower itself. The land is characterizable more accurately as an intersection of extremes: softness and hardness, distance and proximity, and scarcity and density. I have grappled with the multiple narratives used to describe parrot bush, including the master narrative of taxonomy (see Chapter 3). *Budjan*, as the plant is known to the Nyoongar, is also designated by an abundance of ever-evolving technical and colonial names that reflect cultural histories. As a crucible for these heterogeneous meanings, the poem provides a deliberately disjointed, even unnerving, rumination.

Understanding Parrot Bush

beside the rusted out
Survey Corps station
    : budjan in the Dreaming

bolted into limestone occiput
punted by prevailing winds
    : sessilis after Banks

        hypostasis of endurance
        condensed between ocean
        and inner limestone enormity

turret of petals, stamens silky
helter-skelter inside an armamentarium
    : josephia in early taxonomy

        you adapt your downy insides
        softer in hardness, more loving in
        the hardnesses, this land—

a place of beetles' rest ringed
by tough unflinching spikes
    : virile many-flowered dryandra

fair seas west off of Jurien
polygonal interruptions south
    : prickly banksia, coarse to touch

        made bold and brash by abrupt
        inversions of colour and the shock
        that enfolds light-bathed pupils

funnel of mine smoke
lancinating the low heath disarray
    : a man's flora, shaving-brush flower

> at home in erupting psychotropic
> flatness and maddening geometries
> following immeasurable serenities

citrusy bee-stirred nectar
stymieing the pangs of thirst
> : *Europe's holly-leaved dryandra*

>> one tender prod into your
>> silken demesne can never tell
>> how soft you have to be—

a singular solvent thing,
the enchantment of bees,
> : *Parrot Bush, lift your heart to bloom.*

-(in Fremantle Poets 3, edited by Mitchell 2013).

The Phenomenology of Being With

As the Preface described, scientific accounts of South-West plants tend to highlight the world-class biodiversity of the region. I usually give the following one-liner as background to what I do: the South-West corner of Western Australia is a biodiversity hotspot of international significance and one of the most floristically diverse regions in the world (Breeden and Breeden 2010). The "botanical province," including metropolitan Perth, is the only globally recognized biodiversity epicenter in Australia. Isolation is the key; as I have learned, the South-West is an island. The province has long been separated from the rest of southern Australia by the aridity and limestone soils of the inner Nullarbor plains (Hopper 1993). Thus, astonishingly varied and venerable plant communities have evolved through this rare triangulation of stable climate, geographic isolation, and nutrient-starved soils. The relatively flat area of the South-West exhibits a remarkable range of soil types, which give genesis to the notable plant diversity (Corrick and Fuhrer 2002, 13). The biologist and author Barbara York Main (1967, 42) comments:

> There is no landscape more ancient than this anywhere and, because of its age, it has been able, for aeons, to receive and support a fauna and vegetation, limited in variety and density only by the rigorous requirements set by the relatively barren nature of its soils and hazardous climactic conditions.

There are over 8,000 species of indigenous plants, or more than 14 times the number of species found in the entire United Kingdom. Almost 40 percent of the plants in the South-West are endemic—found to occur under natural, uncultivated conditions only within the region (Paczkowska and Chapman 2000).

Yet, factual accounts only go so far in promoting appreciation and even less far in bringing about sensation. I have come to understand landscape as both an aesthetic and scientific abstraction based in visuality. But, for me, plants pierce through appearances with their sensorial particularities. This is where poetry comes in. The plants exact attention, gesture and contact that can make for eco-aesthetic—or at least sense-rich—landscape poetry or, as Chapter 6 posited, *habitat poetry*. Wildflowers impart character to a landscape, but most commonly they are the affectionate objects of sight. The Australian visual artist Gregory Pryor (2005) describes this as the "loaded aesthetic appeal of the flower." Unlike the movement of animals, the subtle motility of plants requires a sensory awareness of our "environment." We need to be vigilant about the conflation of living plants with their representations and with static objects of art. For as Hitchings and Jones (2004, 11) rightly note, "vegetation is something passive in contemporary understanding."

My appreciation of South-West plants—and my environmental aesthetic ideas—confronted an impasse. I needed to infuse the language of flora with the concept and experience of *process*—as both cultural dynamism and ecological movement—to balance the rhetoric of *stasis* that I saw as endemic to a landscape mode of appreciation. Such a mode freezes plants in space and time, creating idealized representatives, not individual organisms or lives but "faceless" members of a collective species. In Heideggerian phenomenology, in contrast, plants epitomize *physis*, a standing forth and going back, a revealing and concealing (Heidegger 1977, 284-317). A process approach to plants counteracts *Ge-stell*, the extraction of a frame in the life of a plant—the strong suit of taxonomic science and the conceptual basis of most botanical art and image-making (Heidegger 1977). On this point, Andrew Feenberg (2005, 30) argues that a plant symbolizes:

> Rootedness in the earth from out of which it emerges. It stands forth from the earth by going back into the earth, sinking its roots into its source. This double movement—standing forth and going back—characterizes the specific motility of living things.

At one level, these poetic interludes attempt to express how plants grow and decay through time in relation to a place, how their sensuousness changes through the seasons and with respect to human contact (see Chapter 1 for a discussion of the Nyoongar six seasons).

Interlude II: Jarrahdale, WA

Considered the first expert on Australian timber, the early Western Australian forester John Ednie-Brown (1899, 10) commented:

> Taken as a whole, there is nothing particularly picturesque about the appearance of a Jarrah tree or forest of these. Indeed, the general effect of the species, en masse, is dull, sombre, and uninteresting to the eye.

Clearly these observations indicate that Ednie-Brown imported perceptual sensibilities to Australian forests, based in appearances and subtended by a managerialist approaches to ecology. However, there are multiple narratives—

Aboriginal, embodied, and poetic—of the jarrah tree (*Eucalyptus marginata*) that come into contact, and at times conflict, with scientific concepts of treeness (see Chapter 3). In Nyoongar belief, the *kaarny* of a recently deceased person would be caught and placed in the burned-out trunk of a jarrah to pacify its restlessness (McCabe 1998, 6). For me, the quintessential act of embodiment is physically entering an old tree and feeling the volatile processes of fire, time, and age that hollowed out its core, leaving a tunnel to the sky. The jarrah forest can be a venue for outdoor sport, an antidote to the city, and a reservoir of visual beauty. Although many things, the jarrah forest can also be a place that commands spiritual respect (Trigger and Mulcock 2005). I posture myself in relation to the corpus of the jarrah, inviting situations in which bodies are brought into proximity. I am not looking *at* the tree, as Ednie-Brown did, but looking at the forest from within the tree (that is, *being with* the tree) as an inversion of the picturesque mode of appreciation.

Inside a Jarrah Tree, A Black Tunnel Reaching Skyward

neatly burned-out innards,
this tree lives on as skin,
still supple and twisting in pleats,
but where did the heart go, and the breast bone
and the heavy, unctuous insides?

the spine endures, knobby column
ripped bare by a magnificent thrust of liquid fire;
but what about the soul,
where is its perch now?

outside, the grass trees don
verdant headdresses over charred land,
and kino sap stamps red
insignias along marri trunks.

have you ever breathed inside a tree
to feel the cool glance of air
where once a molten river ran,
seeing the outside from within?

witchetty grubs or kookaburras might,
clawing skyward towards a portal of light—

But I would not stand here forever.

-(in Fremantle Poets 3, edited by Mitchell 2013).

The Processes of Being With

Process has been my guide. However, there have been two distinct but interrelated processes to consider: my creative writing and research approach in connection to the life events (e.g. flowering) of the plants I write about. My poetry in the field began with physical sensation through gestural exploration of plants, stimulated by a general curiosity about the flora of the region. An initiate from North America, I have needed to learn the languages of South-West Australian places as my second tongue. My body has been the means of exploration and language acquisition. Beginning in 2009, I have recorded my sense impressions in a field notebook that indexes the sensory features of the plants along with the vistas engulfing them. I then turn the stream of consciousness into verse in the field. Metaphors emerge sometimes to express consanguineous relationships between my body and the bodies of plants. Each poem is finalized several months after my experience in the field, creating a gap of time into which can percolate a menagerie of technical facts, namings, and lore along with lingering bodily traces. I also take photographs. The final layout of these photographic poems employs a palimpsest style. The interludes superimpose words over images to suggest the mixing ground of visual impact and sensorial narrative. I also wanted to bring into play the idea of the over-layering of colonial nomenclature onto Indigenous names for plants.

My poetry evokes—and mimics at times—the processes of plants in order to express how they change. I have come to the awareness that the unconscious grouping together of living plants and objects of art rests on the perception of shared stasis. The contemporary plant morphologist Rolf Sattler identifies how this problem has entered scientific understandings of plants. He transfigures the classical binary between stasis and movement by arguing that anatomical structures themselves are processes:

> Structure tends to be considered static, whereas process is dynamic. If we mistake the map for the territory, we conclude that plants consist of structures within which processes occur. On closer inspection we learn, however, that what appears static is in fact also dynamic. (Sattler 1994, 451)

Sattler suggests that the map (for example, the static appearance of form) is not the territory (the complex place of bodily interaction). In contrast to an atomistic view of nature as an aggregation of stable things, process elicits "temporality, historicity, change, and passage as fundamental facts" (Rescher 2000, 3). A being with plants would accommodate "indeterminism, instability, and constant change" (Hagen quoted in Phillips 1999, 579). Correspondingly, an embodied aesthetics of plants and places would reflect these qualities of dynamism. Rather than the sum of its anatomical parts, a plant can be defined by its energetic relations with other creatures in habitats. A sense-rich aesthetics would hold process as an underlying basis for appreciating places through our sensory entanglements.

Importantly, the poet Carl Leggo (2007, 166) reminds us that the term *poetry* comes from the Greek "to make." The root of poetry is *poiēsis*, which may also be defined as "in-becoming." I prefer to define *poiēsis* as "in-the-making" to describe an activity—including one in words—that reveals a process of ongoing engagement with the world, under constant refinement and embodying movement, much like living places themselves. The artist Andy Goldsworthy (quoted in Mabey 2010, 154) points to a comparable creative sympathy for the dynamic interplay between surfaces and depths:

> *Movement, change, light, growth* and *decay* are the lifeblood of nature, the energies I try to tap through my work. I need the *shock of touch*, the resistance of place, materials and weather, the earth as my surface. I want to *get under the surface*. When I work with a leaf, rock, stick, it is not just the material itself, it is opening into the *processes of life within and around it*. [italics added]

Heidegger (1982, 59) refers to poetry as "an experience with language." Poetry is one way of *being with* the world and apprehending its becoming; plants as organisms are in the process of becoming. As a form of enquiry into flora, poetic practice is the undergoing of an experience through the *poiētic* relationship between plants and language. Language traces seed opening, flower decay, the appearance of barrenness after wildflower season, seed germination, flower irruption, and the appearance of fertility after spring rain.

But how would I write these nuances? What ideas would guide the praxis? The concept of *adéquation* offered a promising premise. The ecocritic Sherman Paul (quoted in Tredinnick 2005, 282) describes *adéquation* as "an activity in words that is literally comparable to the thing itself." Paul (quoted in Phillips 1999, 587) further characterizes *adéquation* phenomenologically as "a literary equivalence that respects the thing and lets it stand forth." Yet, adéquation alone seemed too potentially distanced, according to Paul's definitions, for the exploration of intimacies with plants. The Australian postcolonial scholar Paul Carter's invocation of the "middle voice" offers a basis for interconnectivity that fits comfortably with my view of praxis as reflexive engagement over time. Carter (1996, 331) defines the core quality of middle voice as "folding time in the sense that it dissolves the subject-object relation, grounding each in the other, continuously redefining both in terms of each other, so that the two sides exist echoically or simultaneously." The middle voice involves "discontinuous partial selves, or the self as historical process" (White quoted in Carter 1996, 331) or a self that is a "process of continuing becoming," in Smith's terms (quoted in Carter 1996, 331). Dialogical rather than deterministic, the middle voice can encompass and express a more deeply developed sense of self and place vis-à-vis plants and language.

Interlude III: Anstey-Keane Damplands, WA

The Mangles kangaroo paw (*Anigozanthos manglesii*), also known as the red and green kangaroo paw, has been the floral emblem of Western Australia since 1960 (Corrick and Fuhrer 2002, 83) (also see Chapter 4). Nyoongar people have known this iconic South-West plant by its traditional name *kurulbrang*; the tender rhizomes have been traditionally consumed before the emergence of the flower (Hopper 1993, 65). Like other species such the donkey orchid, the kangaroo paw draws its common name zoomorphically. The new flower "capped in green" is concurrent with the onset of spring, the season in which "colour is gestated," but soon the complex blossoms "resign to brown." Through color process, the poem signals the passing of a season within itself. As a macrocosmic unit of change, a season consists of constant microcosmic instances of transformation that herald its overall passage (see Chapter 1). The poem is a haptic meditation on the species.

First Kangaroo Paws

peddling their charms this way—
up briskly from tawny earth
candelabras of crayon red, capped in green,
old tentacles darkening to crimson;

refractions of sunset imprinted in soil
but spiraling back to dust already under
zephyr swoosh and swivel of gum leaves.

the dogs, closer to ground,
imbibe root steams of tepid earth—
stutter and overstep razors
of *Isopogon* and pricks of *Hakea*

leap, pant against barbed bush.
wind-spurred rain skittles
hankering for sun, colour gestates;

spry newbies in variegated cradles,
kangaroo paws crane necks,
resign to brown, shrivel pubescent hope
in glistening perimeters

breathing in Devonian blooms:
bristly hairs ping my nose,
the shimmering shucks off.

-(in Ryan 2012c, Two With Nature)

The Poetics of Being With

The plant-poet nexus surfaces in the literature of most cultures. Poet-botanists from William Wordsworth to D.H. Lawrence wrote at the boundary between the categorization of plants by science and the appreciation of plants by poetry (Mahood 2008) (also see Chapter 6). The botanic-poetic interludes in *Green Sense* (and the examples given in this chapter) extend the rich tradition of plant-based poetry. I have tried to take Henry David Thoreau's ideal to heart. I have aimed to "nail words to their primitive senses" through bodily encounters with flora and the use of words to express those encounters (Thoreau 1862/2007, 29). In contrast, the polemic or analytical slant of most academic writing summons the strengths of the mind rather than the intimacies the natural world "in the between of world and thing." Thus, the interludes are meant to create a kind of chemistry between bodily sensation and intellectual analysis—a style employed by other scholars of arts-based enquiry (for example, Knowles 2001). A short preamble contextualizes each poem and theorizes its arts-based underpinnings. The interludes aim to concretize and materialize aesthetic philosophy in relation to a place, without glossing over my bodily presence or my position as writer, researcher, thinker, or aesthetic subject.

In particular, since the beginning of my place-based project in 2008, I have drawn from the writings of poet Carl Leggo in applying poetic practice to academic knowledge-making. Leggo (2007, 167) suggests that the zones between analytical and creative research are dynamic places, accessible through poetry as "textual spaces that invite and create ways of knowing and becoming in the world." A botanic-poetic interlude opens a space, a pause for rumination, a breath for reflection, and an opportunity to express a sensuous encounter with flora. Whereas Leggo emphasizes the capacity of poetry to balance the logos-privileged discourses of the social sciences, poetic enquiry could be similarly extended to the natural sciences where the attainment of pithy truths underlies specialized ways of knowing the natural world (i.e. scientific disciplines: botany, zoology, entomology, ichthyology). Poetic enquiry into South-West flora summons the "two mingled voices" of literature and science (Hayles 1990, 176-177). The act of writing poetry can foster relationships between plants, people, and places that span the two cultures divide of objective knowledge on the one side and subjective insight on the other.

Despite its social science origin, arts-based research can be broadened to include research into nature and flora. Ardra Cole and J. Gary Knowles (2007, 59) characterizes arts-based research as a "mode and form of qualitative research in the social sciences that is influenced by, but not based in, the arts broadly conceived." Lorri Neilsen (2007, 93-104) uses the term *lyric enquiry* to describe research practices with song-like outcomes. Monica Prendergast (2009, xxxv) defines poetic enquiry as "a form of qualitative research in the social sciences that incorporates poetry in some way as a component of an investigation." Furthermore, Carl Leggo (2007, 168) discusses poetic enquiry *poiētically* as "a way of knowing, being, and becoming in the world." In Leggo's terms, poetry has the ability to trace the being and becoming of people and plants across seemingly static points of reference, such as aesthetic imagery and taxonomic pronouncements. Importantly, for Leggo (2007, 168), poetry expresses "ongoing engagement" with the world beyond the demarcations between the creative arts and social sciences imposed by the academy.

Poetic enquiry can become *geoautobiography*, or the narrative exploration of one's personal history in relation to the story of the land (Porteous 1996, 244). An embodied aesthetics in which the materiality of a place interplays geoautobiographically with the body of the researcher emerges in the work of arts-based researcher Suzanne Thomas (2004) who deploys poetic enquiry methods to develop a research approach to coastal Canadian islands. She writes: "I experience island as an inter-subjective, corporeal encounter—my human body moving in relation to natural bodies of island(s); my body and the bodies of islands in relation to one another, and to the immensity of the sea" (Thomas 2009, 128). In the poem "Prima Materia," Thomas writes of the body of a dead seal: "Ripe flesh, rotting skin | lie transmutable | carrion, offal, microbe, maggot | dissolving body returns to earth" (Thomas 2004, 170). Her body is a sensorium, in which the senses as a whole act as the interface between herself and what she studies. Her senses move between distance and proximity, and her poetic outcomes soften the subject–object assumptions between humans and landscapes, feeling and intellect. The work of Thomas reminds us that bodily decay is an enveloping state of *poiēsis*, poignantly connected to the immediacy of smell.

Expanding the idea of geoautobiography, I liken my approach to *geoautoethnography*, defined as the exploration of one's personal, visceral relationship to places, including the plants of a place. Having qualities of ethnography—which provides the framework for human research in the arts and humanities—such an approach links research choices and behaviors to a place. Poetic enquiry incubates personal memories, cultural histories, quirky anecdotes, taxonomic nomenclatures, metaphorical associations, and emotional insights. Whereas my research could have taken the form of an entirely polemic Foucauldian study of plants, it has evolved a hybridic and ficto-critical orientation, blending analytical and sensory writing. In addition to bodily engagement, I have relied also on taxonomic guides, such as the Department of Environment and Conservation's FloraBase, to understand the plants with which I have been working. As a caveat, my interconnected sense of subjectivity, while a critical working principle for forwarding the concept of *botanical aesthetics* (Chapter 5), also recognizes the improbability of the dissolution of the observing subject and observed object. The historian and philosopher Patrick Curry (2010, 206) characterizes such a position as a "viable middle way, grounded in our embodied, imperfect, unstable, liminal nature."

Interlude IV: John Forrest National Park near Perth, WA

Leggo (2006, 148) advocates a balance between quantitative and qualitative approaches, that we become "present and open sensuously to the whole earth." He advocates sense-rich rumination on the natural world in the form of a:

> Poetics of research in long walks on the dike where I listen to light, smell the line of a heron startled into slow motion by my presence, taste the screech of eagles and hawks, poke with the roots of alders and aspens into the black earth, see the scent of the seasons. (Leggo 2001, 177)

Like Henry David Thoreau, Leggo (2006, 151) enters into the fabric of the world through his *bodily eye*, through corporeal participation: "What I want is to revel in what I am seeing, to see with the whole body, so that my body is rendered alive, is written in the poems." As an example of a poetic narrative of openness to plants, "Sunday Zamia Swagger" plays on the condition zamia staggers, a toxic shock known to develop in cattle grazing on the zamia palm (see Chapter 3). I was intrigued by the dual notions of a plant as a poison to be shunned or destroyed by colonists, but a nutriment to be consumed and fostered by Aboriginal peoples as *djiridji*. I wanted to get close to the villainous species through the embodiment of walking. The result was not only a confrontation with the cultural history of a plant, but a rediscovery of the essential human expressions of gestural curiosity and sense openness.

Sunday Zamia Swagger

by the fire, Sunday morning I imagine *by-yu*
so meander out to the plicae between rolling land
higher to the scarp where the red gums thicken;

a Qantas jet groans, the sun strikes sporadically
under the path of flight through autumn clouds—
from its lonely nook, a dusky roo breaks into
 fricatives;

cross-hatches of wash-outs and dirt tracks
to the bitumen wending west to Swan View—
a scenic vista, lugubrious cars slanted at the edge
a woman with a crew-cut extinguishes a butt
a faceless man slinks into the peace of nothingness

others pass slowly | the way to better things:

an imperturbable hydra, squat black trunk
leaflets stiff as blinds, crisp as piano keys played
forté in one long swipe through seven octaves

tawny cones leaking aloe, striking the nostrils
larghissimo, tessellations of earth acridities
eerily dying back into a rotunda of arachnid legs;

Grey observed "violent fits of vomiting"
Vlamingh, "no distinction between death and us"—
savoring its bready fruits, unsoaked like hazels,
cattle staggered at the poison of the New World,

encased in the sweet flesh of a nut.

-(in Fremantle Poets 3, edited by Mitchell 2013).

Conclusion: The Places of Being With

I have come to feel that sense of place takes shape most for me through close experience of plants in a place over time. When I first came to Perth in 2008, I read the works of the Western Australian ecologist and author of *Sense of Place*, George Seddon. He expressed an initial aversion to South-West plants: "The country was all wrong [...] All the plants scratched your legs. The jarrah was a grotesque parody of a tree, gaunt, misshapen, usually with a few dead limbs, fire-blackened trunk, and barely enough leaves to shade a small ant" (Seddon 1972, xiii-xiv). In *Landprints*, twenty-five years later, Seddon's language blossoms with intimacy and observation. He describes the "profusion of creamy spikes" of *Melaleuca huegelii* and *Acacia rostellifera* "wind-sheared into a dense mound which protects the soil and moulds the landscape" while *Templetonia retusa* "puts out its brick-red pea flowers in spring" (Seddon 1997, 13). Seddon records his identity *in-becoming* in an increasingly familiar place. His growing identity gathered together personal, public, and natural histories and rested on a continuum between bodily aversion (the country was all wrong) and aesthetic revelation (the flowers are all right). As environmental humanities writers have noted, place can be "embodied spatiality," entailing the physical permeability between people and a place (Rose and Robin 2004).

The British nature writer Richard Mabey (2005, 152) comments that "plants are part of what makes a locality, differentiates it, makes an amorphous site into a place, a territory, an address." The interludes presented in the chapter attempt to convey place-consciousness through plants in terms of the relationship between cultural history, ecological meanings, sense invocations, and one's polemic self. To this, Porteous (1996) suggests that sense of place occurs narratively. Moreover, for Susan Griffin (1995, 9), sense of place can be a sacred act of consciousness and embodiment:

> If human consciousness can be rejoined not only with the human body but with the body of earth, what seems incipient in the reunion is the recovery of meaning within existence that will infuse every kind of meeting between self and the universe, even in the most daily acts, with an eros, a palpable love, that is also sacred.

As "Sunday Zamia Swagger" expresses, common gestures of curiosity enable intimate connections to places as sacred "even in the most daily acts." The spatially transformative quality of walking, as an example of a daily act, contests exact geometries with shiftings of shapes and the perception of patterns: circles become lines, spheres are seen as assemblages of squares in places that constantly refuse two-dimensional constructions. Carter (1996, 178-179) uses the term *peripateia* to connote "a measuring-out of consciousness spatially [...] the ground where one walks provides the metre of one's thoughts. The lie of the land, its irregular stresses and glissandi, provides [...] home." Walking is the ultimate act through which my body extends into the world, gestures towards plants, makes contact, and forges intimacies. In Heideggerian terms, this is a movement from a "covetous vision of things" to an intermeshing with the world. It is ultimately a language-based conversion from "the work of the eyes, to the 'work of the heart'" (Heidegger 1971, 136). For me, sense of place importantly involves becoming intimate with the plants of my immediate environment and through the kinds of everyday bodily encounters I have described. At the close of this chapter, I arrive at a metaphor: the delta of plants, processes, and places is nourished by the watery tracings of poetic sensibility and sensory intimacy. And while there can be proximity, there is also *dif-ference*.

Chapter 9: Darwin's Speculative Method: A Poetry of Science, or a Science of Poetry?

Introduction

Specialization was not in the lexicon of Erasmus Darwin (1731–1802): doctor, scientist, poet, inventor, and socialite (Smith and Arnott 2005) (see Figure 11). By profession, he was an unparalleled general physician with a universality of mind that infused his practice of medicine as well as his technical innovation, scientific observation, and poetic vision. As an artist, he achieved notoriety as the sardonic, ponderous bard whose rhythmically perfect couplets embraced the scientific and technological issues of his time. Writing in an instructive and captivating manner, Darwin became the only best-selling scientific poet in English history largely due to his steadfast conviction that poetry should amuse and entertain the public. His poetry, and in particular his choice to recruit science and technology as its subject matter, will be discussed in this chapter. The focus will be on two of Darwin's long poems. The first, *The Botanic Garden*, is divided into Part I, "The Economy of Vegetation" (1791), and Part II, "The Loves of the Plants" (Darwin 1799). The second poem addressed here is the posthumous *The Temple of Nature* (Darwin 1804).

The Speculative Method of Erasmus Darwin

Both poems demonstrate Darwin's talent for writing up scientific findings in verse. *The Botanic Garden* was highly acclaimed during its time and sold widely whereas the success of *The Temple of Nature* was shunned because of Erasmus Darwin's controversial evolutionary ideas. As we will see regarding his evolutionary theory, Darwin's approach to scientific discovery differed radically from the usual, modern picture of the rogue scientist in an antiseptic lab, carefully plotting statistical frequency curves. For Darwin, scientific findings involved a mixture of informed conjecture and imagination. Darwin's approach to poetry hinged on this ability to illumine a scope of subjects with concise couplets supported by imaginative exposition and speculation. Scientific facts and theories interwoven with mythological places and characters constitute his "hypotheses." In the preface to "The Economy of Vegetation," he even apologizes to the reader for putting forth so many conjectures, but asserts that a current of speculation sets in motion a much-needed series of confirmatory experiments (King-Hele 1963, 100). Hence, for Darwin, imagination is the fountainhead of ideas from which scientific discourse springs. Still, why would an esteemed physician write up his scientific "findings," in particular the more controversial contentions, in poetry instead of in more official and professional treatise? In other words, what special advantage did Darwin recognize that poetry presented for swaying the public toward his scientific assertions?

Figure 10. "Flora at Play with Cupid" from *The Botanic Garden* by Erasmus Darwin (1799).
Source: Public Domain, Wikimedia Commons (before 1800).

Before I present three suggestions in response to these questions, it should be mentioned that Darwin maintained a theory of poetry in which prose instructs while poetry amuses. He further believed that poetry, in order to reach the public, should consist largely of visual images: "the Poet writes principally to the eye, the Prose-writer uses more abstracted terms" (Darwin quoted in Logan 1936). Darwin replaced prose abstraction with poetic images by personifying a theoretical scientific account, then having the mythological being behave in a captivating manner. *The Botanic Garden*, for instance, is punctuated with visually sharp mythological episodes, many relating to the Goddess of Botany addressing a humble audience of sylphs, gnomes, and nymphs. Hence, for Darwin, the imaginative yet palpable visual images appropriate to verse provided a forum for speculation whereas the pure abstraction of prose would never hold the same weight in popularizing his scientific findings.

My first suggestion consists of two intertwined points: poetry provided the imaginative medium for Darwin's scientific conjecture and imaginative scenario was the sugary coating that allowed Darwin to feed his scientific and technological ideas to the populace. Section I of this chapter will address this suggestion as it pertains to Darwin's first long poem, *The Botanic Garden*, with emphasis on Part I, "The Economy of Vegetation." *The Botanic Garden* with its complex imaginative structure best exemplifies Darwin's weaving together of scientific observation and innovation with phantasmagoric characters and settings. By entertaining the populace with stunningly crafted verses and amusing preternatural persona, Darwin made his notions palatable to a wide audience and thereby broadcast his scientific speculation (Torrens 2005, 259-272).

Figure 11. "Erasmus Darwin" by Joseph Wright of Derby, 1792, oil on canvas.
Source: Public Domain, Wikimedia Commons.

The second suggestion here is that the versification of science relates to Darwin's theory of organic happiness in which the contentment of a creature enhances its potential to survive. Hence, his scientific poetry with its intention to amuse and delight shows Erasmus working through his own theory; in some way, the couplets are the doctor's attempts to ensure, through organic happiness, his own existence within the scheme of evolution. This point will be expanded in Section I through an examination of Darwin's second long poem, *The Temple of Nature*.

The final suggestion of this paper relates mostly to the first suggestion: poetry serves as a vehicle for his voluminous footnotes and additional notes through which he expounds in more serious tone the innovative scientific concepts put forth in his verse. Thereby, Erasmus could offer the best of both worlds: amusing poetry and didactic prose notes. That the notes often exceed the poems in length implies that Darwin was very confident in the ability of the notes to propound his theoretical ideas while floating under the cover of verse.

Versified Science: Blending Poetic Imagination and Scientific Theory

For Erasmus Darwin, scientific findings involved informed conjecture and hypothesis mixed with imagination. Poetry provided Erasmus a versatile medium in which to freely speculate on scientific matters in a way not possible through formal scientific writing. In fact, after the success of his long poems, Erasmus was often ridiculed with the pejorative term *Darwinizing*, coined by Coleridge to mean "speculating wildly" (King-Hele 1963, 86). Indeed *The Botanic Garden* is unrestrained and honestly impressive in its sheer breadth of wild speculation with all of science as its subject matter. Part I, "The Economy of Vegetation," focuses on physical science and technology with botany making an appearance at the end. The second part, "The Loves of the Plants," reflects its title by detailing the process of fertilization in various classes of plants. In general, the poem anticipates several modern developments in science and technology and illuminates many subjects, including electricity, chemistry, photosynthesis, solar physics, plant physiology, and geomagnetism, which Darwin accurately attributes to the rotation of the planet.

While imagination provided the field for supposition, it also succeeded in communicating Darwin's far-ranging speculation and more grounded hypothesizing in poetic terms that both the general public and literary readers could appreciate. The title of *The Botanic Garden* attracted a gentrified poetry audience to a regime of instruction in science and technology. Most readers would only later realize that very little of the work has to do with plants. Darwin admits his ulterior motive "to inlist Imagination under the banner of Science; and to lead her votaries from the looser analogies, which dress out the imagery of poetry, to the stricter ones, which form the ratiocination [logic] of philosophy" (quoted in King-Hele 1963, 98-99). Imagination brought levity of spirit to an otherwise rote compendium of science. Famous for his wit and robust sarcasm, Darwin further describes the poem as "trivial amusement" (quoted in King-Hele 1968, 193). The appeal of irony to Darwin is evident throughout *The Botanic Garden* where nymphs and dryads dramaticize the bonding of molecules or bemoan the birth of the moon from a gash in the earth's side.

The first two cantos of "The Economy of Vegetation" describe the Earth's origin and its physical features with discourses on geology, chemistry, and technology mixed in. The topics jump between lightning, rainbows, fireballs, comets, steam engines, volcanoes, earthquakes, limestone, porcelain, coal, morasses, mines, minerals, manure, and much more. Around the celestial Goddess of Nature convene nymphs of fire and water, earthly gnomes, and airy sylphs, through whom she will elucidate the origins of natural phenomena. Canto I commences with a discourse to the "nymphs of primeval fire," which explains the origin of the solar system. The planets arose from the Sun and "Bend, as they journey with projectile force | In bright ellipses their reluctant course" (ll. 109–110). Darwin then associates the nymphs of fire with precise descriptions of many fire-related or luminous events. Shooting stars and lightning are linked to the activities of the nymphs: "Ethereal powers! You chase the shooting stars | Or yoke the vollied lightnings to your cars" (ll. 115–116). The nymphs "untwist the sevenfold threads of light" to make rainbows (l. 117) and "fire the arrowy throne of rising Morn" to create morning and evening sky colors (l. 119). Darwin alludes to the hydrogen composition of the outermost atmosphere "Where lighter gases, circumfused on high | Form the vast concave of exterior sky" (ll. 123–124). The nymphs are responsible for the aurora, which "Dart from the North on pale electric streams | Fringing Night's sable robe with transient beams" (ll. 129–130). In the eighteenth century, auroral lights were classified as terrestrial phenomena, but Darwin was perspicacious in recognizing that the aurora belonged to the outermost layer of the atmosphere (King-Hele 1963, 103). The fervid catalogue goes on with planets, fixed stars, volcanoes, and phosphoric lights. Darwin then postulates that flying machines powered with steam or another explosive material will be used in war:

> Soon shall thy arm, UNCONQUER'D STEAM! Afar
> Drag the slow barge, or drive the rapid car;
> Or on wide-waving wings expanded bear
> The flying-chariot through the fields of air.
> Fair crews triumphant, leaning from above
> Shall wave their fluttering kerchiefs as they move;
> Or warrior-bands alarm the gaping crowd,
> And armies shrink beneath the shadowy cloud. (ll. 289-296)

The first canto finishes with the nymphs of fire exiting in a firework exhibition. During this grand exeunt, Darwin shows his humorous side by describing electricity as an aphrodisiac: "Through her fine limbs the mimic lightnings dart | And flames innocuous eddy round her heart" (ll. 349-353).

The second canto is directed to the gnomes of earth and proposes a version of the Earth's creation. Here Darwin demonstrates his keen sense of geological history. The gnomes are guardians of the natal Earth, born recently from the fiery sun. Darwin details the evolution of the oceans from the Earth's "vaporous air, condensed by cold" and characterizes gravitational pull as "fierce attraction with relentless force" that "bent the reluctant wanderer to its course" (ll. 11-20). Darwin speculates that the moon originated in the Pacific Ocean, a proposal later argued for mathematically by his great-grandson George Darwin (King-Hele 1963, 103). The elder Darwin fancies that the gnomes were aghast when the moon came into existence "from her wounded side | Where now the South-Sea heaves its waste of tide" (ll. 77-78).

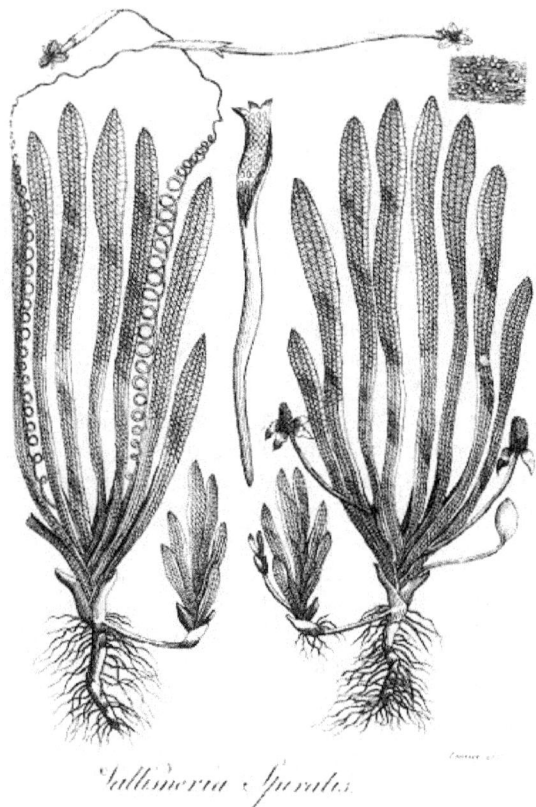

Figure 12. *Vallisneria spiralis* (1789–1791) from *The Botanic Garden* by Erasmus Darwin (1799).
Source: Public Domain, Wikimedia Commons.

After a tumultuous early period, the history of earth smoothed out. Here, Darwin attends to the geological and atmospheric composition of the planet. He first explains the process by which limestone forms from shells and how bogs yield iron salts. Darwin contends that most rocks form from organic remains combined with existing minerals and he includes notes on coal, flint, clay, and calcareous earth (King-Hele 1963, 108). He specifically discusses three regions of the atmosphere. The first, which is the site of cloud-formation, goes up about four miles and corresponds to the troposphere in modern atmospheric science. He estimated the second region to extend thirty-five miles above the first layer. Lighter gases consisting mostly of hydrogen dominated the most distant region beyond thirty-five miles. For his time, Darwin was remarkably accurate, although we now know that hydrogen does not become the dominant gas until about 2000 miles out (King-Hele 1963, 107).

The third canto of *The Economy of Vegetation* addresses various aquatic subjects including clouds, rain, rivers, canals, pumps, and geysers. The aquatic nymphs supervise the water cycle of the planet by leading "the winged vapours up the aerial arch" (ll. 12-13), then down once more as rivers, seas, and fluids within organisms. Darwin is usually accurate in his discourses on lower atmospheric occurrences such as evaporation, clouds, and winds. He revels in the omnipresence of water from the rivers beneath snow-clad summits and the tumult of tropical monsoons to glacial geysers and hot springs. The nymphs are aware of the powerful chemical bonds between the "pure air" of oxygen and the "flaming Gas" of hydrogen that constitute water:

> Nymphs! Your bright squadrons watch with chemic eyes
> The cold-elastic vapours, as they rise;
> With playful force arrest them as they pass,
> And to pure air betroth the flaming Gas.
> Round their translucent forms at once they fling
> Their rapturous arms, with silver bosoms cling. (ll. 201-206)

This is a prime example of Darwin combining rote scientific facts and imagination in the form of spry nymphs provoking loving embrace between hydrogen and oxygen with stable water the result of their union. The image, though preposterous, is quite effective. The poet then turns his attention, as he so freely does, from science to technology by discussing water pumps for extinguishing fires or draining swamps. The nymphs instructed mortals "to pierce the secret caves | Of humid earth, and lift her ponderous waves" (ll. 345-346). Darwin again shows his dexterity in setting scientific fact in imaginative contexts by switching from the nymphs to a treatment of pump pressure dynamics:

> Bade with quick stroke the sliding piston bear
> The viewless columns of incumbent air;
> Press'd by the incumbent air the floods below,
> Through opening valves in foaming torrents flow. (ll. 346-350)

By expressing his grasp of scientific and technological issues through verses such as these, Darwin adds weight to the fanciful mechanics of *The Economy of Vegetation*. Through such a fusion of science and imagination, the poet has achieved a compelling vehicle for substantiating his scientific observations with a mere trimming of empirical backing, if any at all.

Canto IV begins with monsoons, fogs, barometers, and submarines, but, alas, vegetation makes an appearance towards the end with a treatment of the anatomy and physiology of seeds, flowers, and leaves. The sylphs of the air stimulate wind currents in the atmosphere and make plants transpire oxygen. In more taciturn moods, the sylphs bring fog and tornadoes. Darwin entreats them to explain how to master winds:

> Oh, Sylphs! disclose in this inquiring age
> One Golden Secret to some favour'd sage;
> Grant the charm'd talisman, the chain, that binds,
> Or guides the changeful pinions of wind! (ll. 307-310)

Afterall, the sylphs showed Torricelli and Boyle "How up exhausted tubes bright currents flow...And with the changeful moment fall and rise" (ll. 130 & 134), so they must be able to disclose the pattern behind the fickleness of the winds. Turning to technology, the identification of oxygen suggested to Darwin the feasibility of extended submarine travel: "The diving castles, roof'd with spheric glass | Ribb'd with strong oak, and barr'd with bolts of brass" (ll. 197-198). In this final canto, Darwin charts the course for Part II, "The Loves of the Plants," by propounding the general principle of vegetable growth and reproduction. *The doctor* here emphasizes the vital role of light in stimulating oxygen to escape from the leaves of plants to the atmosphere; *the poet* imagines that oxygen is infatuated with light.

Whereas "The Economy of Vegetation" is a broad survey of science and technology, "The Loves of the Plants" is a tightly focused discourse on plant reproductive mechanisms in which Darwin discloses the romantic lives of almost one hundred species. Although this second part brought Darwin fame after its 1789 publication, the skillful couplets and mythic setting barely disguise the abiding monotony of the subject matter (King-Hele 1963, 110). Consequently, Darwin humanizes each plant according to its color, form, habitat, or method of fertilization. Arguing that plants experience sensations of love, he substituted English equivalents for the Latin names in order to craft human metaphors that entertained readers who could appreciate an endearing love story. As a result of shrewd entrenchment of plant physiology in a morass of romantic tomfoolery, the average reader of the late 1700s probably had a broad grasp of botany that would be humbling by our modern standard.

Darwin begins the second part by explaining the Linnaean plant classes as distinguished by the character of the reproductive components (see Figure 12). He then humanizes some of the plant reproductive characteristics. Indian reed is virtuous because one male (stamen) and one female (pistil) in each flower are engaged in "their nuptial vow" (l. 48). The stargrass, conversely, shares two virgins (two pistils) with one male and *Collinsonia* is impious in pairing two males (two stamens) to one female. However, with Dyer's broom for which "ten fond brothers woo the haughty maid" (l. 58) polygamy has fully erupted. Amidst the hedonism, however, Darwin stays true to scientific fact when he alludes to "Two knights before thy fragrant altar bend | Adored Melissa! And two squires attend" (l. 65) to account for the two higher (two knights) and two lower (two squires) stamens arranged on the broom species, *Genista melissa*. Darwin continues to lighten the botanical tedium with comic scenarios. With regard to *Lychnis*, for instance, the five females "rise above the petals, as if looking abroad for their distant husbands" (l. 144). Unlike Dyer's broom, the sensitive mimosa is prudish: "Weak with nice sense, the chaste Mimosa stands | From each rude touch withdraws her timid hands" (ll. 301-302). One may suspect that Darwin was questioning the human practice of monogamy by implying that it was rare in nature. In another sense, the numerous lascivious episodes of the poem presumably enhanced sales (King-Hele 1963, 115).

The final three cantos of "The Loves of the Plants" continue the botanical catalogue but with more frequent and longer flights of reverie not solely of the romantic kind. Although science and technology are not the dominant subjects in final three cantos, Darwin does offer some of the technical insight that so thoroughly characterizes "The Economy of Vegetation." The flying seeds of thistle recall ballooning: "So on the shoreless air the intrepid Gaul | Launch'd the vast concave of his buoyant ball' (ll. 25-26). The cotton plant with its "vegetable wool" prompts complex descriptions of spinning machines (ll. 34). Darwin then goes on to extol the Peruvian bark and foxglove for their capacities to restore health (l. 378). The third canto introduces the nightshade, laurel, and upas tree as somber characters in a rather melancholy canto. The final canto concludes the poem on a lighter note. A gnat that carries pollen from stamen to the "secret cave" of the pistil fertilizes a wild fig (l. 428). Darwin concludes the canto with Adonis, in which many males and females live together in the same flower: "A hundred virgins join a hundred swains | And fond Adonis leads the sprightly trains" (ll. 423-424).

Evolution and Darwin's Theory of Organic Happiness

Darwin's second long poem, *The Temple of Nature*, was published after his death, in 1803. In this poem Darwin propounded a theory of evolution that was more complete than any earlier model and came quite close to the modern theory propounded by his grandson, Charles. Erasmus placed less emphasis than earlier models on the transmission of acquired characteristics, and he astutely estimated the temporal scale of evolution as millions of years. The elder Darwin, however, failed to produce the empirical evidence that his grandson would furnish in becoming one of the world's most esteemed scientists (King-Hele 1968, 182).

If Darwin lacked a mathematical vehicle to give credence to a theory of evolution, he found an imagination-based one in poetry. Personifications of nature, love, and death preside over the poem. An impressive portrait of the Temple of Nature unfolds. The Priestess of Nature, Urania, who resides in the Temple of Nature, narrates most of the story. Through this second long poem, Darwin asserts the idea that life originated as microscopic specks in primeval seas and gradually, under evolutionary pressures, developed its present forms. As is the case for his first long poem, *The Temple of Nature* conveys his scientific theory to the public through carefully crafted couplets. Unlike the rather pell-mell *The Botanic Garden*, however, Darwin's second long poem provides a coherent chronological structure based on evolutionary progression. *The Temple of Nature* had all the makings of a second best seller but was frozen in its tracks by Darwin's contention that evolutionary processes go on without assistance from any deity. The public was especially sensitive at a time when they were relying on "the Deity" to help them resist Napoleon's encroachment on the southern coast of England (King-Hele 1968, 152).

Despite its contentiousness, *The Temple of Nature* presents a picture of evolution over fifty years before it came to be accepted by scientists. The first canto shows the origin of life and the evolution of aquatic organisms to land creatures. The main theme of the cantos is the slow evolution of microscopic life in primeval oceans with the subsequent transition to amphibious forms and finally to terrestrial life. Darwin is certain that complex life developed over enormous periods of time from the simplest forms to the most sophisticated, including humanity:

Organic life beneath the shoreless waves
Was born and nurs'd in Ocean's pearly caves;

> First forms minute, unseen by spheric glass,
> Move on the mud, or pierce the watery mass;
> These, as successive generations bloom,
> New powers acquire, and larger limbs assume; (ll. 295-300)

Darwin tacitly parallels evolutionary development to the progress of an embryo from the "waves" of the amniotic fluid to post-natal land as a "dry inhabitant of air:"

> Thus in the womb the nascent infant laves
> Its natant form in the circumfluent waves;
> With perforated heart unbreathing swims;
>
> [...]
>
> Awakes and stretches all its recent limbs...
> Gives to the passing gale his curling hair,
> And steps a dry inhabitant of air. (ll 389-393, 399-400)

These insinuations were probably at the core of the public's rejection of *The Temple of Nature*:

The second canto deals with asexual, hermaphroditic, and, finally, sexual reproduction. Darwin contends that asexual reproduction came first: "The Reproductions of the living Ens | Form sires to sons, unknown to sex, commence" (ll. 63-64). The inadequacy of asexual reproduction became evident as:

> The feeble births acquired diseases chase
> Till Death extinguish the degenerate race...
> Or till, amended by connubial powers
> Rise seedling progenies from sexual flowers. (ll. 165-166, 175-176)

Hymen then announces Cupid and Psyche, "the Deities of Sexual Love" (see Figure 10). Darwin, in keeping with his stated intention to amuse and entertain, follows this up with many comical examples of married bliss or strife in the natural world: "The Lion-King forgets his savage pride | And courts with playful paws his tawny bride" (ll. 357-358).

The third canto follows the evolution of the mind from its origin as a nexus of nerves to its present complexity in humans. Urania tells the Muse how the human faculty of reason developed. Darwin emphasizes two physical attributes that made possible human evolutionary progress, hand and eye: "Nerved with fine touch above the bestial throngs | The hand, first gift of Heaven! to man belongs" (ll. 121-122). Sight is described as "the mute language of the touch" (l. 144). Darwin, quintessentially jumping among topics, then offers a philosophy of art and science, which he thinks can be explained in terms of imitation:

> Hence to clear images of form belong
> The sculptor's statue, and the poet's song,
> The painter's landscape, and the builder's plan,

And imitation marks the mind of man. (ll. 331-334)

His idea is that we merely imitate what we see or what others have done; such imitations, in the hands of an artist or scientist, occasionally achieve meaning.

The fourth canto, "Of Good and Evil," describes, in grim terms, the struggle for existence and the survival of the fittest, then relishes the pleasures of life and outlines the organic theory of happiness. The beginning of the canto includes images of the wolf tearing the helpless lamb, the eagle rending the innocuous dove, the lamb and dove grazing the young herb or seedling, and the owl diving for the glowing worm, "Who climbs the green stem, and slays the sleeping flower" (ll. 17-28). The world is "one great Slaughter-house" (l. 66). Balancing out the troubles of humans, the triumphs of science include the sublime concepts of Newton and the practical insights of Archimedes. Darwin postulates that these innovations in science are the precursors to skyscrapers, traffic, and piped water:

Bid raised in air the ponderous structure stand,
Or pour obedient rivers through the land
With cars unnumber'd crowd the living streets,
Or people oceans with triumphant fleets. (ll. 315-318)

Despite the inevitable disastrous effects of excessive overbreeding, Darwin proposed a theory of organic happiness as part of his evolutionary ideas. The theory of evolution became a scientific hypothesis that served as a basis of Erasmus Darwin's philosophy of life, including all organic life (King-Hele 1963, 90). He suggests that as a product of the system of reproduction and death, "Happiness survives" (ll. 452-453); amidst the potential strife of overpopulation, a fundamental happiness will prevail. He saw hopeful patterns controlling the perpetual massacre. Darwin's theory of organic happiness, which most fully is expressed in this final canto, suggests that, since the beginning, species have been enhanced by evolution. The survival of the fittest has been the prolongation of the best species or members of species. Darwin also believed that life is an essentially enjoyable process and that the happier organism has a better chance of surviving the struggle for existence. As we all have a common microscopic ancestor, we should view the animals, insects, and plants as our relatives: "His brother emmets, and his sister worms" (ll. 427-428). While alive, an organism contributes to the sum total of happiness; when it dies the animal or plant brings happiness to other organisms.

The elder Darwin presents an optimistic take on evolution, a process more typically associated with the unsympathetic substitution of one species for another in the perpetual battle to survive. The *summon bonum* of evolution, as expressed toward the end of *The Temple of Nature*, is the happiness and interrelation of all life. His consistent declarations that his poetry should be taken as light amusement could reflect his penultimate realization that more complex life, though established through suffering, affords more substantial happiness. Hence, he chose to write up his evolutionary conjectures in poetry as an expression of his personal contribution to organic happiness. His poetry, through its levity and whimsicality, represents his whittling of a niche in the evolutionary scheme, so to speak.

Sneaking in Scientific Exposition: The Footnotes Scheme

The final idea relating to why Erasmus Darwin chose to write up his scientific findings in poetry ties in with the first suggestion that poetry allowed the author to broadcast his empirically unsubstantiated claims to a wide audience. Darwin used voluminous explanatory notes to supplement the verse. These notes were obviously written in prose form, which according to Darwin's theory of poetry is best suited to didacticism. The tone of the notes is ponderous, as one would expect of a scientific treatise; there are no nymphs, sylphs, or dryads, no romantic innuendoes, no triumphant goddesses of nature. The sheer thesis-length of the notes suggests that these were not mere supplements to the verse, intended to clarify abstract scientific concepts. Unlike the poetry, the prose notes are serious and Darwin may have thought that they were necessary to advance his theories even further. Inserting them under the cover of poetry was a tactical decision that reflects the choice to not submit them as separate scientific papers.

The footnotes to "The Economy of Vegetation" are longer than the poem itself and cover almost every branch of science and technology. To supplement the footnotes, there are about 50,000 words of additional notes on such diverse subjects as meteors, primary colors, colored clouds, comets, the sun's rays, heat, the steam engine, electricity, oil on water, and glaciers. One of the "Notes" is a 9000-word discourse on winds, including a meteorological journal. The essay speculates about the worldwide circulation of air. Darwin demonstrates full awareness of the role of vertical motions and horizontal winds. The notes on the Earth's interior spanned 15,000 words and include a summary of the geological history of the planet. Darwin's geological expositions in *The Botanic Garden*, which form a 15,000 word treatise, contrast to all the faulty catastrophic theories of the time. Moreover, Darwin's secular geological discourse excels his contemporary theorists who sought the divine wisdom in their ideas.

The footnotes to *The Temple of Nature* speculate with equal breadth on all branches of science. In the first note, Darwin discusses spontaneous vitality. He believes that microscopic life can in certain instances arise from non-living matter within a few days. The experiments on which he bases his hypothesis were later discredited and spontaneous generation became an abandoned issue. However, as is the case for many of Darwin's other speculations, the idea has been revitalized recently. We even see the beginnings of a germ theory:

> Thus one grain of variolous matter, inserted by inoculation, shall in about seven days stimulate the system into unnatural action; which in about seven days more produces ten thousand times the quantity of a similar material thrown out on the skin in the postules! (38, *Temple of Nature*, "Additional Notes").

Darwin appreciated the importance of microscopic "animalcules" and the exponential nature of successive reproduction. Further in the "Additional Notes" of the poem, Darwin offers a two-fluid theory of electricity and magnetism. Similarly, in propounding his theory of evolution, Darwin offers numerous, extensive notes to amplify the argument:

> After islands or continents were raised above the primeval ocean, great numbers of the most simple animals would attempt to seek food at the edges of the shores of the new land, and might thence become amphibious...Those situated on dry land, and immersed in dry air, may gradually acquire new powers to preserve their existence... (1. 327, "Note").

It is clear that Darwin's notes were not intended to be typical footnotes, in the sense of minor adjunctive explanations. The prose notes may have been designed to offer an alternative reading to those who were suspicious of the intermingling of mythology and science. Darwin may have been trying to maintain his scientific credibility by including rather formal scientific treatises scattered throughout the whimsical cantos.

However, as I have tried to suggest in this chapter, Darwin probably knew that many of his scientific theories lacked adequate empirical bulwarking. Part of Darwin's wide-ranging treatment of scientific findings in imaginary contexts reflects the fact that his forward thinking outpaced the development of explanatory techniques. A different view is that Darwin was simply a scholar of "breadth" rather than "depth." Evolutionary theory, once again, is a quintessential example. Erasmus could only speculate about the progression of simple animals seeking "food at the edges of the shores of the new lakes" who then became amphibious and finally creatures of "dry air [with] new powers to preserve their existence." Charles Darwin later found in the Galapagos and elsewhere the much-needed evidence to substantiate these conjectures. Aware of the shortcomings of his speculative theories, accurate nonetheless, Darwin chose to use poetry as a disguise.

Erasmus Darwin spread his extraordinary talents across the spectrum of human knowledge as many Renaissance figures do. The consequence: a mind-boggling range of subjects converging in his poetry but a conspicuous scientific inadequacy characterizing many. Much of his scientific theorizing has been overlooked, or, more tragically, successive scientists have been credited for the theories. For example, Erasmus Darwin was correct in recognizing that the aurora was an electrical phenomenon of the outermost atmosphere above thirty-five miles from the surface of the Earth. Yet, the auroral measurements in the International Polar Year of 1882 were fruitless because the stations were situated according to the assumption that the aurora was only five miles high (King-Hele 1968, 190).

Conclusion: Speculating about Darwin's Versification of Science

The doctor's motive to convey his scientific findings to the public through poetry was most likely a mixture of the three suggestions. Instructing readers in science through poetry related to his life's work as a general physician, as one at service to the public. Through this simultaneous entertainment and instruction of the people, Darwin manifested his organic happiness in keeping with the theory of evolution. In a different and less light-hearted sense, Darwin probably knew he lacked the palpable evidence to adequately propel his theories to respectability. Thus, on one level, Darwin was entirely serious about setting science to verse and including exceedingly long prose notes since the poetic artifice could amuse the reader as a cover for his lack of empirical depth. The simple answer can only be as complex as Darwin himself and the natural world he poeticized.

References

Adamson, Joni, and Catriona Sandilands. 2013. "Vegetal Ecocriticism: The Question of "The Plant." Preconference Panel, Changing Nature: ASLE Tenth Biennial Conference, Kansas. Accessed May 15, 2013. http://asle.ku.edu/Preconference/pdf/Vegetal_Ecocriticism_The_Question_of_ThePlant.pdf

Adorno, Theodor. 1984. *Aesthetic Theory*. Translated by Christian Lenhardt. London: Routledge & Kegan Paul.

Alexander, Alan. 1979. "Nuytsia Floribunda." In *Wide Domain: Western Australian Themes & Images*, edited by Bruce Bennett and William Grono. London: Angus & Robertson.

Anderson, Kat. 1993. "Native Californians as Ancient and Contemporary Cultivators." In *Before the Wilderness: Environmental Management by Native Californians*, edited by Thomas Blackburn and Kat Anderson, 151-174. Menlo Park: Ballena Press.

Armstrong, Francis. 1979. "Manners and Habits of the Aborigines of Western Australia, From Information Collected by Mr F. Armstrong, Interpreter, 1836." In *Nyungar - The People: Aboriginal Customs in the Southwest of Australia*, edited by Neville Green, 186-206. North Perth: Creative Research.

Australian Bureau of Meteorology. 2008a. *A Century of Science and Service: The Australian Bureau of Meteorology 1908-2008*. Melbourne: Bureau of Meteorology

Australian Bureau of Meteorology. 2008b. *Climate of Australia*. Melbourne: Bureau of Meteorology.

Australian Bureau of Meteorology. 2012a. "Indigenous Weather Knowledge." Accessed January 21, 2012. http://www.bom.gov.au/iwk/.

Australian Bureau of Meteorology. 2012b. *Reconciliation Action Plan 2012-2015*. Melbourne: Bureau of Meteorology.

Australian Bureau of Meteorology. 2013. "Australian Weather Calendar." Accessed January 28, 2013. http://www.bom.gov.au/calendar/annual/seasons.shtml.

Aveni, Anthony. 1990. *Empires of Time: Calendars, Clocks, and Cultures*. London: I.B Tauris.

Bain, M.A. 1975. *Ancient Landmarks: A Social and Economic History of the Victoria District of Western Australia 1839-1894*. Nedlands: University of Western Australia Press.

Barad, Karen. 2010. "Quantum Entanglements and Hauntological Relations of Inheritance: Dis/continuities, SpaceTimeEnfoldings, and Justice-to-Come." *Derrida Today* 3 (2): 240-268.

Barker, Chris. 2008. *Cultural Studies: Theory and Practice*. 3 ed. London: Sage.

Barrell, John. 1972. *The Idea of Landscape and the Sense of Place 1730-1840: An Approach to the Poetry of John Clare*. Cambridge: Cambridge University Press.

Barrett, Russell, and Eng Pin Tay. 2005. *Perth Plants: A Field Guide to the Bushland and Coastal Flora of Kings Park and Bold Park, Perth, Western Australia*. West Perth: Botanic Gardens and Parks Authority.

Bates, Daisy. 1985. *The Native Tribes of Western Australia*. Edited by Isobel White. Canberra: National Library of Australia.

Bates, Daisy. 1992. *Aboriginal Perth and Bibbulmun Biographies and Legends*. Victoria Park: Hesperian Press.

Beard, John, and John Pate. 1984. "Foreword." In *Kwongan: Plant Life of the Sandplain*, edited by John Beard and John Pate. Nedlands: University of Western Australia.

Beckey, Fred. 1993. *Mount McKinley: Icy Crown of North America*. Seattle: Mountaineers.

Bede. 1999. *Bede: The Reckoning of Time*. Translated by Faith Wallis. Liverpool: Liverpool University Press.

Bender, Barbara. 2002. "Time and Landscape." *Current Anthropology* 43 (S4): 103-112.

Beresford, Quentin, Hugo Bekle, Harry Phillips, and Jane Mulcock. 2001. *The Salinity Crisis: Landscape, Communities and Politics*. Crawley: University of Western Australia Press.

Berleant, Arnold. 2005. *Aesthetics and Environment: Variations on a Theme*. Burlington: Ashgate Publishing.

Bindon, Peter, and Trevor Walley. 1992. "Hunters and Gatherers." *Landscope* 8 (1): 28-35.

Blackburn, Bonnie, and Leofranc Holford-Strevens. 1999. *The Oxford Companion to the Year: An Exploration of Calendar Customs and Time-Reckoning*. Oxford: Oxford University Press.

Borst, Arno. 1993. *The Ordering of Time: From the Ancient Computus to the Modern Computer*. Cambridge: Polity Press.

Bowman, David, and Tom Vigilante. 2001. "Conflagrations: The Culture, Ecology, and Politics of Landscape Burning in the Northern Kimberley." *Ngoonjook* 20: 38-45.

Brady, Emily. 2003. *Aesthetics of the Natural Environment*. Edinburgh: Edinburgh University Press.

Brain, Tracy. 2001. *The Other Sylvia Plath*. Harlow: Pearson Education.

Breeden, Stanley, and Kaisa Breeden. 2010. *Wildflower Country: Discovering Biodiversity in Australia's Southwest*. Fremantle: Fremantle Press.

Bropho, Robert. 1998. "Interview 17: Robert Bropho. Swan Valley Aboriginal Community, Circle of Elders. 9:30am, 17:06:98." In *Nyoongar Views on Logging Old Growth Forests*, edited by Timothy McCabe, 30-32. West Perth: Wilderness Society.

Brown, Andrew, Andrew Batty, Mark Brundrett, and Kingsley Dixon. 2003. "Underground Orchid (*Rhizanthella gardneri*) Interim Recovery Plan 2003-2008." Accessed October 2, 2010. http://www.environment.gov.au/biodiversity/threatened/publications/rec overy/r-gardneri/background.html.

Bryson, J. Scott. 2002. *Ecopoetry: A Critical Introduction*. Salt Lake City: The University of Utah Press.

Burnett, Viv. 2008. *Establishing Perennial Pastures: The Foundation for Sustainable Organic Farming Systems*. Barton: RIRDC.

Butscher, Edward. 1976. *Sylvia Plath: Method and Madness*. New York: The Seabury Press.

Caddy, Caroline. 1989. *Beach Plastic*. Fremantle: Fremantle Arts Centre Press.
Carlson, Allen. 1993. "Appreciating Art and Appreciating Nature." In *Landscape, Natural Beauty and the Arts*, edited by Salim Kemal and Ivan Gaskell, 199-227. Cambridge: Cambridge University Press.
Carlson, Allen. 2000. *Aesthetics and the Environment: The Appreciation of Nature, Art, and Architecture*. New York: Routledge.
Carr, D.J., and S.G.M. Carr. 1981. "The Botany of the First Australians." In *People and Plants in Australia*, edited by D.J. Carr and S.G.M. Carr, 3-44. Sydney: Academic Press.
Carter, Paul. 1996. *The Lie of the Land*. London: Faber & Faber.
Choate, Alec. 1978. *Gifts Upon the Water*. Fremantle: Fremantle Arts Centre Press.
Choate, Alec. 1986. *A Marking of Fire*. Fremantle: Fremantle Arts Centre Press.
Choate, Alec. 1995. *Mind in Need of a Desert*. Fremantle: Fremantle Arts Centre Press.
Christensen, Laird. 2002. "The Pragmatic Mysticism of Mary Oliver." In *Ecopoetry: A Critical Introduction*, edited by J Bryson, 135-152. Salt Lake City: The University of Utah Press.
Clarke, Philip. 2007. *Aboriginal People and their Plants*. Dural Delivery Centre: Rosenberg.
Clarke, Philip. 2009. "Australian Aboriginal Ethnometeorology and Seasonal Calendars." *History and Anthropology* 20 (2): 79-106.
Coates, Peter. 2006. *American Perceptions of Immigrant and Invasive Species: Strangers on the Land*. Berkeley: University of California Press.
Colbung, Ken. 1998. "Interview 30: Ken Colbung. Elder. Joondalup. 9:30am. 10:07:98." In *Nyoongar Views on Logging Old Growth Forests*, edited by Timothy McCabe, 51-56. West Perth: Wilderness Society.
Cole, Ardra, and J. Gary Knowles. 2007. "Arts-Informed Research." In *Handbook of the Arts in Qualitative Research: Perspectives, Methodologies, Examples, and Issues*, edited by J. Gary Knowles and Ardra Cole, 55-70. Thousand Oaks: Sage.
Collard, Dorothy. 1998. "Interview 19: Dorothy Collard. Actress and Great Grandmother." In *Nyoongar Views on Logging Old Growth Forests*, edited by Timothy McCabe, 32-35. West Perth: Wilderness Society.
Collins, Kevin, Kathy Collins, and Alex George. 2008. *Banksias*. Melbourne: Blooming Books.
Corrick, Margaret G., and Bruce Fuhrer. 2002. *Wildflowers of Southern Western Australia*. Noble Park: The Five Mile Press.
Cosgrove, Denis. 1998. *Social Formation and Symbolic Landscape*. Madison: University of Wisconsin Press.
Cotton, C.M. 1996. *Ethnobotany: Principles and Applications*. Chichester: John Wiley & Sons.
Crary, Jonathan. 1990. *Techniques of the Observer: On Vision and Modernity in the Nineteenth Century*. Cambridge: MIT Press.
Cross, Nigel. 2011. *Design Thinking: Understanding How Designers Think and Work*. Oxford: Berg.

CSIRO. 2012. "Indigenous Seasonal Indicators and Climate Change." Accessed December 17, 2012. http://www.csiro.au/Organisation-Structure/Divisions/Ecosystem-Sciences/Ngadju.aspx.
Cunningham, Irene. 2005. *The Land of Flowers: An Australian Environment on the Brink*. Caringbah: Otford Press.
Curry, Patrick. 2010. "Grizzly Man and the Spiritual Life." *Journal for the Study of Religion, Nature and Culture* 4 (3): 206-219. doi: 10.1558/jsrnc.v413.206.
Darwin, Erasmus. 1799. *The Botanic Garden, A Poem. In Two Parts*. 4 ed. London: J. Johnson, St. Paul's Church-Yard.
Darwin, Erasmus. 1804. *The Temple of Nature; or, The Origin of Society: A Poem with Philosophical Notes*. Baltimore: John W. Butler, and Bonsal & Niles.
Daw, Brad, Trevor Walley, and Greg Keighery. 1997. *Bush Tucker Plants of the South-West*. Kensington: Department of Environment and Conservation.
Day, Gordon. 1953. "The Indian as an Ecological Factor in the Northeastern Forest." *Ecology* 34 (2): 329-346.
Dean, Bradley P. 2000. "Introduction." In *Wild fruits: Thoreau's Rediscovered Last Manuscript*, edited by Bradley P Dean. New York: W.W. Norton & Company.
Denes, Alexandra. 2012. "Acquiring the Tools for Safeguarding Intangible Heritage: Lessons from an ICH Field School in Lamphun, Thailand." In *Safeguarding Intangible Cultural Heritage*, edited by Michelle Stefano, Peter Davis, and Gerard Corsane, 165-176. Woodbridge: Boydell & Brewer.
Department of Environment and Conservation. 2012. "FloraBase: The Western Australian Flora." Accessed August 19, 2012. http://florabase.dec.wa.gov.au/.
Doggett, L.E. 1992. "Calendars." In *Explanatory Supplement to the Astronomical Almanac*, edited by P. Kenneth Seidelmann, 575-608. Mill Valley: University Science Books.
Don, David. 1835. "*Anigozanthos manglesii*: Mr. Mangles's *Anigozanthos*." In *The British Flower Garden*, edited by Robert Sweet, 265-266. London: James Ridgway.
Donaldson, Mike. 1996a. "The End of Time?: Aboriginal Temporality and the British Invasion of Australia." *Time & Society* 5 (2): 187-207.
Donaldson, Mike. 1996b. *Taking Our Time: Remaking the Temporal Order*. Nedlands: University of Western Australia Press.
Duncan, Katie. 2011, April 1."Sprinting Into Sprummer." In *Australian Geographic. Academic OneFile*. Accessed January 15, 2013.
Eaton, Marcia. 1998. "Fact and Fiction in the Aesthetic Appreciation of Nature." *The Journal of Aesthetics and Art Criticism* 56 (2): 149-156.
Edith Cowan University. 2013. "ECU and BOM Launch Nyoongar Weather Calendar." Accessed February 5, 2013. http://www.ecu.edu.au/news/latest-news/2012/12/ecu-and-bom-launch-nyoongar-weather-calendar.
Ednie-Brown, John. 1899. *The Forests of Western Australia and Their Development, with Plan and Illustrations*. Perth: Perth Printing Works.

Elkin, Adolphus Peter. 1943. *The Australian Aborigines: How to Understand Them*. Sydney: Angus & Robertson.
Elliott, Brian. 1967. *The Landscape of Australian Poetry*. Melbourne: F.W. Cheshire.
Erickson, Dorothy. 2009. *A Joy Forever: The Story of Kings Park & Botanic Garden*. West Perth: Botanic Gardens & Parks Authority.
Farr, Douglas. 2012. *Sustainable Urbanism: Urban Design with Nature*. Hoboken: John Wiley & Sons.
Feenberg, Andrew. 2005. *Heidegger and Marcuse: The Catastrophe and Redemption of History*. New York: Routledge.
Feeney, Denis. 2007. *Caesar's Calendar: Ancient Time and the Beginnings of History*. Berkeley: University of California Press.
Felstiner, John. 2009. *Can Poetry Save the Earth? A Field Guide to Nature Poems*. New Haven: Yale University Press.
Flinders, Matthew. 1814. *A Voyage to Terra Australis*. Accessed January 23, 2013. http://books.google.com.
Franklin, R.W. (ed.). 1999. *The Poems of Emily Dickinson: Reading Edition*. Cambridge: The Belknap Press of Harvard University Press.
Fredregill, Ernest. 1970. *1000 Years: A Julian/Gregorian Perpetual Calendar, A.D. 1100 to A.D. 2099*. New York: Exposition Press.
Gadd, Geoff. 2000. "Heterotrophic Solubilization of Metal-Bearing Minerals by Fungi." In *Environmental Mineralogy: Microbial Interactions, Anthropogenic Influences, Contaminated Land and Waste Management*, edited by J. Cotter-Howells, L. Campbell, E. Valsami-Jones and M. Batchelder, 57-75. Middlesex: Mineralogical Society of Great Britain & Ireland.
Gell, Alfred. 1992. *The Anthropology of Time: Cultural Constructions of Temporal Maps and Images*. Oxford: Berg.
George, Alex. 2002a. *The Long Dry: Bush Colours of Summer and Autumn in South-Western Australia*. Kardinya: Four Gables Press.
George, Alex. 2002b. "The South-western Australian Flora in Autumn: 2001 Presidential Address." *Journal of the Royal Society of Western Australia* 85: 1-15.
Giblett, Rod. 2011. *People and Places of Nature and Culture*. Bristol: Intellect.
Gifford, Terry. 1995. *Green Voices*. Manchester: Manchester University Press.
Girardet, Herbert. 2008. *Cities, People, Planet: Urban Development and Climate Change*. 2 ed. Chichester: John Wiley.
Glendinning, Simon. 2007. *In the Name of Phenomenology*. London: Routledge.
Graham, Duncan. 1990. *Nyoongar Bush Tucker with Ken Colbung* [VHS]. Mount Lawley: The Institute of Applied Aboriginal Studies.
Grainger, Percy. 1985. *The Farthest North of Humanness: Letters of Percy Grainger 1901-14*. Translated by Kay Dreyfus. South Melbourne: The Macmillan Company.
Grant, Jaime. 1994. "LANSDOWN, Andrew." In *The Oxford Companion to Twentieth-century Poetry in English*, edited by Ian Hamilton. New York: Oxford University Press.
Green, Neville. 1984. *Broken Spears: Aborigines and Europeans in the Southwest of Australia*. Cottesloe: Focus Education Services.

Grey, George. 1841a. *Journals of Two Expeditions of Discovery in North-West and Western Australia During the Years 1837, 38, and 39*. Vol. 1. London: T. and W. Boone.

Grey, George. 1841b. *Journals of Two Expeditions of Discovery in North-west and Western Australia During the Years 1837, 38, and 39*. Vol. 2. London: T. and W. Boone.

Griffin, David. 1994. *Fungal Physiology*. 2 ed. New York: Wiley-Liss.

Griffin, Susan. 1995. *The Eros of Everyday Life: Essays on Ecology, Gender and Society*. New York: Doubleday.

Groser, Thomas. 1927. *The Lure of the Golden West: Experiences and Adventures in a Bush Brotherhood of Western Australia: Early Problems and Conquests: All About Group Settlements of the West: The Land of Sunshine and Opportunity*. London: Alexander-Ouseley.

Hall, Matthew. 2009. "Plant Autonomy and Human-Plant Ethics." *Environmental Ethics* 31: 169-181.

Hallam, Sylvia. 1989. "Plant Usage and Management in Southwest Australian Aboriginal Societies." In *Foraging and Farming: The Evolution of Plant Exploitation*, edited by David Harris and Gordon Hillman, 136-151. London: Unwin Hyman.

Hallam, Sylvia J. 1975. *Fire and Hearth: A Study of Aboriginal Usage and European Usurpation in South-Western Australia*. Canberra: Australian Institute of Aboriginal Studies.

Haraway, Donna. 2008. *When Species Meet*. Minneapolis: University of Minnesota Press.

Harper, Douglas. 2012. "habitat." *Online Etymology Dictionary*. Accessed May 16, 2012. http://www.etymonline.com/index.php?term=habitat.

Harper, Douglas. 2013. "phenology." *Online Etymology Dictionary*. Accessed February 5, 2013. http://www.etymonline.com/index.php?term=phenology.

Harper, Douglas. 2013. "respect." *Online Etymology Dictionary*. Accessed February 5, 2013. http://www.etymonline.com/index.php?allowed_in_frame=0&search=respect&searchmode=none

Haskell, Dennis. 2000. "Tradition and Questioning: *The Silo* as Pastoral Symphony." In *Fairly Obsessive: Essays on the Works of John Kinsella*, edited by Rod Mengham and Glen Phillips, 89-102. Fremantle: Fremantle Arts Centre Press.

Haskell, Dennis, and Hilary Fraser. 1989a. "Alec Choate." In *Wordhord: Contemporary Western Australian Poetry*, edited by Dennis Haskell and Hilary Fraser. Fremantle: Fremantle Arts Centre Press.

Haskell, Dennis, and Hilary Fraser. 1989b. "Andrew Lansdown." In *Wordhord: Contemporary Western Australian Poetry*, edited by Dennis Haskell and Hilary Fraser. Fremantle: Fremantle Arts Centre Press.

Hassell, Ethel 1975. *My Dusky Friends*. Dalkeith: C.W. Hassell.

Hawkins, Aaron. 1751. *The Gregorian and Julian Calendars, or, The New and Old Stiles, Arithmetically Explained*. London: M. Cooper. Accessed January 29, 2013. http://find.galegroup.com/mome/infomark.do?&source=gale&prodId=M

OME&userGroupName=uwa&tabID=T001&docId=U3600934148&type=multipage&contentSet=MOMEArticles&version=1.0&docLevel=FASCIMILE.

Hayles, N. Katherine. 1990. *Chaos Bound: Orderly Disorder in Contemporary Literature and Science*. Ithaca: Cornell University.

Heidegger, Martin. 1971. *Poetry, Language, Thought*. Translated by Albert Hofstadter. New York, NY: Harper Perennial.

Heidegger, Martin. 1977. "The Question Concerning Technology." In *Martin Heidegger: Basic Writings*, edited by D. Krell, 284-317. New York: Harper & Row.

Heidegger, Martin. 1982. *On the Way to Language*. Translated by Peter Hertz. San Francisco: Harper & Row.

Heidegger, Martin. 2009. "The Age of the World Picture." In *The Heidegger Reader*, edited by Gunter Figal, 207-223. Bloomington: Indiana University Press.

Helms, Eleanor. 2008. "Language and Responsibility: The Possibilities and Problems of Poetic Thinking for Environmental Philosophy." *Environmental Philosophy* 5 (1): 23-36.

Hewett, Dorothy. 2001. *Halfway Up the Mountain*. Fremantle: Fremantle Arts Centre Press.

Heyd, Thomas. 2001. "Aesthetic Appreciation and the Many Stories About Nature." *British Journal of Aesthetics* 41 (2): 125-137.

Hill, Mike. 1998. "Interview 7: Mike Hill. RFA Nyoongar Action Group Chairperson. Public Meeting." In *Nyoongar Views on Logging Old Growth Forests*, edited by Timothy McCabe, 18-20. West Perth: Wilderness Society.

Hitchings, Russell. 2003. "People, Plants and Performance: On Actor Network Theory and the Material Pleasures of the Private Garden." *Social & Cultural Geography* 4 (1): 99-113.

Hitchings, Russell, and Verity Jones. 2004. "Living With Plants and the Exploration of Botanical Encounter within Human Geographic Research Practice." *Ethics, Place & Environment* 7 (1 & 2): 3-18.

Hoffman, Bruce, and Timothy Gallaher. 2007. "Importance Indices in Ethnobotany." *Ethnobotany Research & Applications* 5: 201-218.

Holbrook, David. 1976. *Sylvia Plath: Poetry and Existence*. London: University of London; The Athlone Press.

Holford-Strevens, Leofranc. 2005. *History of Time: A Very Short Introduction*. Oxford: Oxford University Press.

Hopper, Stephen. 1993. *Kangaroo Paws and Catspaws: A Natural History and Field Guide*. Como: Department of Conservation and Land Management.

Hopper, Stephen. 1998. "An Australian Perspective on Plant Conservation Biology Practice." In *Conservation Biology for the Coming Decade*, edited by Peggy Fiedler and Peter Kareiva, 255-278. New York: Chapman Hall.

Hopper, Stephen. 2010. "Nuytsia floribunda." *Curtis's Botanical Magazine* 26 (4): 333-368. doi: 10.1111/j.1467-8748.2009.01671.x.

Hostetler, Mark. 2012. *The Green Leap: A Primer for Conserving Biodiversity in Subdivision Development*. Berkeley: University of California Press.
Hsu, Elisabeth. 2010. *Plants, Health and Healing: On the Interface of Ethnobotany and Medical Anthropology*. New York: Berghahn Books.
Huggan, Graham, and Helen Tiffin. 2010. *Postcolonial Ecocriticism: Literature, Animals, Environment*. New York: Routledge.
Hunn, Eugene. 1990. *Nch'i-Wana "The Big River:" Mid-Columbia Indians and their Land*. Seattle: University of Washington Press.
Isaacs, Jennifer. 1989. *Bush Food: Aboriginal Food and Herbal Medicine*. Chatswood: New Holland Publishers.
Isar, Yudhishthir Raj, and Helmut Anheier (eds.). 2011. *Heritage, Memory & Identity*. Los Angeles: Sage.
Jay, Martin. 1993. *Downcast Eyes: The Denigration of Vision in Twentieth-Century French Thought*. Berkeley: University of California Press.
Kant, Immanuel. 1978. *Anthropology from a Pragmatic Point of View*. Translated by Victor Lyle Dowdell. Carbondale: Southern Illinois University Press.
Keatley, Marie, and Irene Hudson. 2010. "Introduction and Overview." In *Phenological Research: Methods for Environmental and Climate Change Analysis*, edited by Marie Keatley and Irene Hudson, 1-22. Dordrecht: Springer.
Kew. 2005. "Plant Cultures: Exploring Plants and People." Accessed April 1 2013. http://www.kew.org/plant-cultures/index.html.
King-Hele, Desmond. 1963. *Erasmus Darwin*. New York: St. Martin's Press.
King-Hele, Desmond. 1968. *The Essential Writings of Erasmus Darwin*. London: MacGibbon & Kee Ltd.
Kinsella, John. 1997. *Poems 1980-1994*. Fremantle: Fremantle Arts Centre Press.
Kinsella, John. 2005. *The New Arcadia*. New York: Norton.
Kinsella, John. 2009. "Biographical Notes." In *The Penguin Anthology of Australian Poetry*, edited by John Kinsella, 395-435. Camberwell: Penguin Group.
Kitayama, Kanehiro. 2012. *Co-benefits of Sustainable Forestry: Ecological Studies of a Certified Bornean Rain Forest*. Dordrecht: Springer.
Knapp, Sonja. 2010. *Plant Biodiversity in Urbanized Areas*. Wiesbaden: Springer Fachmedien.
Knickerbocker, Scott. 2012. *Ecopoetics: The Language of Nature, the Nature of Language*. Amherst: University of Massachusetts Press.
Knowles, J. Gary. 2001. "Writing Place, Wondering Pedagogy." In *The Art of Writing Inquiry*, edited by Lorri Neilsen, Ardra Cole, and J. Gary Knowles, 89-99. Halifax: Backalong Books.
Kurongkurl Katitjin Centre. 2011. "Wongi Nyoongar - Talking Nyoongar." *Our Place: Official Newsletter of Kurongkurl Katitjin Centre for Indigenous Australian Education and Research*. Makaru Edition (Jun/Jul): 5. Accessed May 24, 2013. http://www.ecu.edu.au/__data/assets/pdf_file/0011/249293/Our-Place-Newsletter_Makuru-11_Edition_WEB.pdf.
Lansdown, Andrew. 1979. *Homecoming*. Fremantle: Fremantle Arts Centre Press.
Lansdown, Andrew. 2009. *Birds in Mind: Australian Nature Poems*. Capalaba: Wombat Books.

Latour, Bruno. 1999. "Circulating Reference: Sampling the Soil in the Amazon Forest." In *Pandora's Hope: Essays on the Reality of Science Studies*, 24-79. Cambridge: Harvard University Press.

Latz, Peter. 1995. *Bushfires and Bushtucker: Aboriginal Plant Use in Central Australia*. Alice Springs: IAD Press.

Lawrence, D.H., and Mollie Skinner. 1924. *The Boy in the Bush*. Edited by Paul Eggert. Cambridge: Cambridge University Press.

Leggo, Carl. 2001. "Research as Poetic Rumination: Twenty-Six Ways of Listening to Light." In *The Art of Writing Inquiry*, edited by Lorri Neilsen, Ardra Cole, and J. Gary Knowles, 173-195. Halifax: Backalong Books.

Leggo, Carl. 2006. "Attending to Winter: A Poetics of Research." In *Spirituality, Ethnography, and Teaching: Stories from Within*, edited by W Ashton and D Denton, 140-155. New York: Peter Lang.

Leggo, Carl. 2007. "Astonishing Silence: Knowing in Poetry." In *Handbook of the Arts in Qualitative Research: Perspectives, Methodologies, Examples, and Issues*, edited by J. Gary Knowles and Ardra Cole, 165-174. Thousand Oaks: Sage Publications.

Levin, David. 1993. "Decline and Fall: Ocularcentrism in Heidegger's Reading of the History of Metaphysics." In *Modernity and the Hegemony of Vision*, edited by David Levin, 186-217. Los Angeles: University of California Press.

Lewis, Henry. 1978a. *Fires of Spring*. [VHS]. Alberta: University of Alberta, 32 minutes.

Lewis, Henry. 1978b. "Traditional Uses of Fire by Indians in Northern Alberta." *Current Anthropology* 19 (2): 401-402.

Lewis, Henry. 1989. "Ecological and Technological Knowledge of Fire: Aborigines Versus Park Rangers in Northern Australia." *American Anthropologist* 91 (4): 940-961.

Leyew, Zelealem. 2011. *Wild Plant Nomenclature and Traditional Botanical Knowledge Among Three Ethnolinguistic Groups in Northwestern Ethiopia*. Addis Ababa: Organisation for Social Science Research in Eastern and Southern Africa.

Lindley, John. 1840. *A Sketch of the Vegetation of the Swan River Colony*. London: James Ridgway.

Logan, James Venable. 1936. *The Poetry and Aesthetics of Erasmus Darwin*. Princeton: Princeton University Press.

Longino, Helen. 1997. "Cognitive and Non-Cognitive Values in Science: Rethinking the Dichotomy." In *Feminism, Science, and the Philosophy of Science*, edited by Lynn Hankinson Nelson, 39-58. Dordrecht: Kluwer Academic Publishers.

Lunt, Ian, and John Morgan. 2002. "The Role of Fire Regimes in Temperate Lowland Grasslands of South-Eastern Australia." In *A Flammable Australia: The Fire Regimes and Biodiversity of a Continent*, edited by Ross Bradstock, Jann Williams, and Malcolm Gill, 177-198. Cambridge: Cambridge University Press.

Mabey, Richard. 2010. *A Brush With Nature: 25 Years of Personal Reflections on the Natural World*. London: BBC Books.

MacDonald, Lindsay. 2012. *Digital Heritage*. Hoboken: Taylor & Francis.
Mahood, Molly. 2008. *The Poet as Botanist*. Cambridge: Cambridge University Press.
Main, Barbara York. 1967. *Between Wodjil and Tor*. Brisbane and Perth: The Jacaranda Press and Landfall Press.
Marchant, N.G., J.R. Wheeler, B.L. Rye, E.M. Bennett, N.S. Lander, and T.D. Macfarlane. 1987. *Flora of the Perth Region: Part One*. Perth: Western Australian Herbarium, Department of Agriculture.
Marder, Michael. 2013. *Plant-Thinking: A Philosophy of Vegetal Life*. New York: Columbia University Press.
Marsden-Smedley, Jon. 2000. "Fire Management in Tasmania's Wilderness World Heritage Area: Ecosystem Restoration Using Indigenous-Style Regime Fires?" *Ecological Management & Restoration* 1 (3): 195-203.
Martin, Gary. 2004. *Ethnobotany: A Methods Manual*. London: Earthscan.
Maughan, Graham. 1996. *Second Nature: Building Forests in West Africa's Savannas*. [VHS]. Haywards Heath: Cyrus Productions, 41 minutes.
McCabe, Timothy. 1998. "Nyoongar Kaarny Nyoongar Spirituality." In *Nyoongar Views on Logging Old Growth Forests*, edited by Timothy McCabe, 6. West Perth: Wilderness Society.
McKinney, Michael. 2005. "Urbanization as a Major Cause of Biotic Homogenization." *Biological Conservation* 127 (3): 247-260.
McKusick, James. 2000. *Green Writing: Romanticism and Ecology*. New York: St. Martin's Press.
Meagher, Sara J. 1974. "The Food Resources of the Aborigines of the South-West of Western Australia." *W.A. Museum Records* 3 (1).
Meinel, Christoph. 2011. *Design Thinking: Understand - Improve - Apply*. Dordrecht: Springer.
Merleau-Ponty, Maurice. 2012. *Phenomenology of Perception*. Translated by Colin Smith. New York: Routledge.
Methuen, Charlotte. 2008. *Science and Theology in the Reformation: Studies in Theological Interpretation and Astronomical Observation in Sixteenth-Century Germany*. London: T & T Clarke.
Michels, Agnes Kirsopp. 1967. *The Calendar of the Roman Republic*. Princeton: Princeton University Press.
Millar, J. 2002. *Mycological Studies*. Ottawa: Canada Council for the Arts.
Millett, Janet. 1872. *An Australian Parsonage or, The Settler and the Savage in Western Australia*. Nedlands: University of Western Australia Press.
Mitchell, Scott-Patrick (ed.). 2013. *Fremantle Poets 3: Performance Poets*. Fremantle: Fremantle Press.
Money, Nicholas. 2011. *Mushroom*. Oxford: Oxford University Press.
Moore, George Fletcher. 1846. *Descriptive Vocabulary of the Language in Common Use Amongst the Aborigines of Western Australia*. Nedlands: University of Western Australia Press. Reprinted in 1978.
Moore, George Fletcher. 1884. *Diary of Ten Years Eventful Life of an Early Settler in Western Australia and Also A Descriptive Vocabulary of the Language of the Aborigines*. Nedlands: University of Western Australia Press. Reprinted in 1978.

Moore, George Fletcher. 2006. *The Millendon Memoirs: George Fletcher Moore's Western Australian Diaries and Letters, 1830-1841*. Edited by J.M.R. Cameron. Carlisle: Hesperian Press.
Morton, Tim. 2010. "The Dark Ecology of Elegy." In *The Oxford Handbook of the Elegy*, edited by Karen Weisman, 251-271. Oxford: Oxford University Press.
Moyal, Ann. 1986. *A Bright & Savage Land: The Science of a New Continent - Australia - Where All Things Were "Queer and Opposite"*. Ringwood: Penguin.
Mules, Warwick. 2008. "Open Country: Towards a Material Environmental Aesthetics." *Continuum* 22 (2): 201-212.
Mules, Warwick. 2014. *With Nature: Nature Philosophy as Poetics through Schelling, Heidegger, Benjamin and Nancy*. Bristol: Intellect.
Murdoch University. 2003. "A Nyungar Interpretive History of the Use of Boodjar in the Vicinity of Murdoch University, Perth, Western Australia." Accessed August 19, 2012. http://wwwmcc.murdoch.edu.au/multimedia/nyungar/menu9.htm.
Nancy, J-L. 2000. *Being Singular Plural*. Translated by R. Richardson. Stanford: Stanford University Press.
Nannup, Noel, and David Deeley. 2006. "Rainfall and Water as Cultural Drivers." In *1st National Hydropolis Conference*. Perth: Stormwater Industry Association.
Neilsen, Lorri. 2007. "Lyric Enquiry." In *Handbook of the Arts in Qualitative Research: Perspectives, Methodologies, Examples, and Issues*, edited by J. Gary Knowles and Ardra Cole, 93-104. Thousand Oaks: Sage Publications.
Nelson, Richard. 1983. *Make Prayers to the Raven: A Koyokon View of the Northern Forest*. Chicago: University of Chicago Press.
Nikulinsky, Philippa, and Stephen Hopper. 1999. *Life on the Rocks: The Art of Survival*. Fremantle: Fremantle Arts Centre Press.
Nikulinsky, Philippa, and Stephen Hopper. 2005. *Soul of the Desert*. Fremantle: Fremantle Arts Centre Press.
Nind, Scott. 1979. "Description of the Natives of King George's Sound (Swan River Colony) and Adjoining Country." In *Nyungar, the People: Aboriginal Customs in the Southwest of Australia*, edited by Neville Green, 14-55. North Perth: Creative Research in association with Mt. Lawley College.
North, Marianne. 1892. *Recollections of a Happy Life*. London: Macmillan.
Northover, Joe. 1998. "Interview 21: Joe Northover. Police Officer. Collie. 4:30pm. 28:06:98." In *Nyoongar Views on Logging Old Growth Forests*, edited by Timothy McCabe, 35-37. West Perth: Wilderness Society.
Oliver, Mary. 1992. *New and Selected Poems*. Boston: Beacon Press.
Paczkowska, Grazyna, and Alex R. Chapman. 2000. *The Western Australian Flora: A Descriptive Catalogue*. Perth: Wildflower Society of Western Australia (Inc.), the Western Australian Herbarium, CALM and the Botanic Gardens & Parks Authority.
Pardo-de-Santayana, Manuel, Andrea Pieroni, and Rajindra Puri. 2010. "The Ethnobotany of Europe, Past and Present." In *Ethnobotany in the New*

Europe: People, Health and Wild Plant Resources, edited by Manuel Pardo-de-Santayana, Andrea Pieroni, and Rajindra Puri, 1-15. New York: Berghahn Books.

Perth Biodiversity Project. n.d. *What is the Perth Biodiversity Project?* West Perth: Perth Biodiversity Project.

Phillips, Dana. 1999. "Ecocriticism, Literary Theory, and the Truth of Ecology." *New Literary History* 30 (3): 599.

Phillips, Glen. 1988. *Sacrificing the Leaves*. Thailand: 10th World Congress of Poets.

Plath, Sylvia. 1967. *The Colossus*. London: Faber and Faber.

Plattner, Hasso. 2012. *Design Thinking Research: Measuring Performance in Context*. Dordrecht: Springer.

Porteous, J. Douglas. 1989. *Planned to Death: The Annihilation of a Place Called Howdendyke*. Manchester: Manchester University Press.

Porteous, J. Douglas. 1996. *Environmental Aesthetics: Ideas, Politics and Planning*. London: Routledge.

Pouliot, Alison, and Tom May. 2010. "The Third 'F' - Fungi in Australian Biodiversity Conservation: Actions, Issues and Initiatives." *Mycologia Balcanica* 7: 27-34.

Prendergast, Monica. 2009. "Introduction: The Phenomena of Poetry in Research." In *Poetic Inquiry: Vibrant Voices in the Social Sciences*, edited by Monica Prendergast, Carl Leggo, and Pauline Sameshima, xix-xlii. Rotterdam: Sense Publishers.

Prober, Suzanne, Michael O'Connor, and Fiona Walsh. 2011. "Australian Aboriginal Peoples' Seasonal Knowledges: A Potential Basis for Shared Understanding in Environmental Management." *Ecology and Society* 16 (2): 1-16, http://www.ecologyandsociety.org/vol16/iss2/art12/.

Pryor, Gregory. 2005. "Black Death: Species Extinction in WA." *Artlink* 25 (4), http://www.artlink.com.au/articles.cfm?id=2227.

Reid, Neil. 2012. *Local Food Systems in Old Industrial Regions: Concepts, Spatial Context, and Local Practices*. Farnham: Ashgate Publishing.

Rescher, N. 2000. *Process Philosophy: A Survey of Basic Issues*. Pittsburgh: University of Pittsburgh Press.

Richards, E.G. 1999. *Mapping Time: The Calendar and Its History*. Oxford: Oxford University Press.

Rodaway, Paul. 2002. *Sensuous Geographies: Body, Sense and Place*. Abingdon: Taylor & Francis.

Roehl, Renée, and Kelly Chadwick (eds.). 2010. *Decomposition: An Anthology of Fungi-Inspired Poems*. Sandpoint: Lost Horse Press.

Rolston, Holmes. 1995. "Does Aesthetic Appreciation of Landscapes Need to be Science-Based?" *British Journal of Aesthetics* 35 (4): 374-385.

Rose, Deborah Bird. 1992. *Dingo Makes Us Human: Life and Land in an Aboriginal Culture*. Cambridge: Cambridge University Press.

Rose, Deborah Bird. 1996. *Nourishing Terrains: Australian Aboriginal Views of Landscape and Wilderness*. Canberra: Australian Heritage Commission.

Rose, Deborah, and Libby Robin. 2004. "The Ecological Humanities in Action: An Invitation." *Australian Humanities Review* 31-32, http://www.australianhumanitiesreview.org/archive/Issue-April-2004/rose.html.

Rusack, Eleanor May, Joe Dortch, Ken Hayward, Michael Renton, Mathias Boer, and Pauline Grierson. 2011. "The Role of *Habitus* in the Maintenance of Traditional Noongar Plant Knowledge in Southwest Western Australia." *Human Ecology* 39 (5): 673-682.

Russell-Smith, Jeremy. 2013. "Indigenous Fire Practices in Western Arnhem Land: Lessons for Today." Accessed February 19, 2013. http://www.savanna.cdu.edu.au/downloads/indfire.pdf.

Russell, Emily. 1983. "Indian-Set Fires in the Forests of the Northeastern United States." *Ecology* 64 (1): 78-88.

Ryan, John. 2012a. *Green Sense: The Aesthetics of Plants, Place and Language.* Oxford: TrueHeart Press.

Ryan, John. 2012b. "The Six Seasons: Shifting Australian Nature Writing Towards Ecological Time and Embodied Temporality." *Transformations* 21, http://www.transformationsjournal.org/journal/issue_21/article_01.shtml

Ryan, John. 2012c. *Two With Nature: The Botanical Art of Ellen Hickman and the Botanical Poetry of John Ryan.* Fremantle: Fremantle Press.

Ryan, John. 2012d. "Which to Become? Encountering Fungi in Australian Poetry." *Rupkatha Journal on Interdisciplinary Studies in Humanities* 4 (2): 132-143.

Ryan, John. 2013. *Unbraided Lines: Essays in Environmental Thinking and Writing.* Champaign: Common Ground.

Ryan, Tracy. 2002. *Hothouse.* Fremantle: Fremantle Arts Centre Press.

Saito, Yuriko. 1998. "Appreciating Nature On Its Own Terms." *Environmental Ethics* 20: 135-149.

Salvado, Dom Rosendo 1977. *The Salvado Memoirs: Historical Memoirs of Australia and Particularly of the Benedictine Mission of New Norcia and of the Habits and Customs of the Australian Natives.* Translated by E.J. Stormon. Nedlands: University of Western Australia Press.

Sattler, Rolf. 1994. "Homology, Homeosis, and Process Morphology in Plants." In *Homology: The Hierarchical Basis of Comparative Biology*, edited by B.K. Hall, 423-475. San Diego: Academic Press.

Schachtel, Ernest. 1984. *Metamorphosis: On the Development of Affect, Perception, Attention, and Memory.* New York: Da Capo Press.

Schwartz, Mark. 2003. "Introduction." In *Phenology: An Integrative Environmental Science*, edited by Mark Schwartz, 3-8. Dordrecht: Kluwer Academic Publishers.

Schwartz, Susan. n.d. "Lecture 19: Alaska Geology and Denali National Park." Accessed November 5, 2011. http://ic.ucsc.edu/~susans/eart3/Lectures/lecture20.html.

Seddon, George. 1972. *Sense of Place: A Response to an Environment.* Perth: University of Western Australia Press.

Seddon, George. 1997. *Landprints: Reflections on place and landscape.* Cambridge: Cambridge University Press.

Seddon, George. 2005. *The Old Country: Australian Landscapes, Plants and People.* Cambridge: Cambridge University Press.

Serres, Michel. 2008. *The Five Senses: A Philosophy of Mingled Bodies*. London: Continuum.
Shapiro, Kenneth, and Margo DeMello. 2010. "The State of Human-Animal Studies." *Society & Animals* 18 (3): 2-17.
Shearer, B.L, C.E. Crane, S. Barrett, and A. Cochraine. 2007. "*Phytophthora cinnamomi* Invasion, a Major Threatening Process to Conservation of Flora Diversity in the South-west Botanical Province of Western Australia." *Australian Journal of Botany* 55 (3): 225-238.
Shenton, Mrs. Edward. 1927. "Reminiscences of Perth, 1830-1840." *The Western Australian Historical Society Journal and Proceedings* 1 (1): 1-4.
Skinner, Mary Louisa. 1972. *The Fifth Sparrow: An Autobiography*. Sydney: Sydney University Press.
Smith, C.U.M., and Robert Arnott. 2005. "Introduction." In *The Genius of Erasmus Darwin*, edited by C.U.M. Smith and Robert Arnott, 1-4. Aldershot: Ashgate Publishing.
Sobel, David. 2008. *Childhood and Nature: Design Principles for Educators*. Portland: Stenhouse Publishers.
Sorensen, Marie Louise Stig, and John Carman (eds.) 2009. *Heritage Studies: Methods and Approaches*. Milton Park: Routledge.
South West Aboriginal Land & Sea Council. 2009. *"It's Still in My Heart, This Is My Country:" The Single Noongar Claim History*. Crawley: University of Western Australia Press.
Stasiuk, Glen, and Ash Sillifant. 2005. *Noongar of the Beelier or Swan River, Australia* [DVD]. Murdoch: Kulbardi Productions.
Stefano, Michelle, Peter Davis, and Gerard Corsane. 2012. "Touching the Intangible: An Introduction." In *Safeguarding Intangible Cultural Heritage*, edited by Michelle Stefano, Peter Davis, and Gerard Corsane, 1-5. Woodbridge: Boydell & Brewer.
Steffen, Will, Andrew Burbidge, Lesley Hughes, Roger Kitching, David Lindenmayer, Warren Musgrave, Mark Stafford Smith, and Patricia A. Werner. 2009. *Australia's Biodiversity and Climate Change*. Collingwood: CSIRO Publishing.
Stein, Rachel. 1997. *Shifting the Ground: American Women Writers' Revisions of Nature, Gender, and Race*. Charlottesville: The University of Virginia Press.
Summers, Lise. 2011. "Wildflower Season: The Development of Native Flora Protection Legislation in Western Australia, 1911-1975." *Studies in Western Australian History* 27: 31-44.
The Perth Gazette. 1833. September 7. "The Natives: Interesting Interview." *The Perth Gazette and Western Australian Journal (WA: 1833-1847)*, http://nla.gov.au/nla.news-article641889.
The University of Western Australia. 2010. "Alec Choate (1915-2010)." Accessed October 28, 2010. http://www.news.uwa.edu.au/201008042732/alec-choate-1915-2010.
The West Australian. 1924. September 19. "Our Wildflowers." *The West Australian* 6.
The West Australian. 1924. September 26. "Our Wildflowers." *The West Australian* 6.

The West Australian. 1928. September 7. "Our Wildflowers: Swan River Myrtle." *The West Australian* 7.
The Western Mail. 1922. March 2. "Western Australian Trees." *The Western Mail* 30.
The Western Mail. 1929. February 7. "Stinkwood and Swish-bush." *The Western Mail* 38.
Thomas, Suzanne. 2004. *Of Earth and Flesh and Bones and Breath: Landscapes of Embodiment and Moments of Re-enactment*. Halifax: Backalong Books.
Thomas, Suzanne. 2009. "Nissopoesis: Visuality and Aesthetics in Poetic Inquiry." In *Poetic Inquiry: Vibrant Voices in the Social Sciences*, edited by Monica Prendergast, Carl Leggo, and P. Sameshima, 127-132. Rotterdam: Sense Publishing.
Thompson, Paul. 2010. *The Agrarian Vision: Sustainability and Environmental Ethics*. Lexington: The University Press of Kentucky.
Thoreau, Henry David. 2007. *Walking*. LaVergne: Filiquarian Publishing. Originally published in 1862.
Thoreau, Henry David. 2010. *Wild Apples*. LaVergne: Filiquarian Publishing. Originally published in 1862.
Thoreau, Henry David. 1962. "The Maine Woods." In *Thoreau: Walden and Other Writings*, edited by J. Krutch. New York: Bantam Books.
Thoreau, Henry David. 1993. *Faith in a Seed: The Dispersion of Seeds and Other Late Natural History Writings*. Washington, DC: Island Press.
Thoreau, Henry David. 2000. *Wild Fruits: Thoreau's Rediscovered Last Manuscript*, edited by Bradley Dean. New York: W.W. Norton & Company.
Tilbrook, Lois. 1983. *Nyungar Tradition: Glimpses of Aborigines of South-Western Australia 1829-1914*. Nedlands: University of Western Australia Press.
Tilley, Christopher. 1994. *A Phenomenology of Landscape: Places, Paths and Monuments*. Oxford: Berg.
Tilley, Christopher. 2010. "Outline of a Phenomenological Perspective." In *Interpreting Landscapes: Geologies, Topographies, Identities*, 25-40. Walnut Creek: Left Coast Press.
Torrens, H.S. 2005. "Erasmus Darwin's Contribution to the Geological Sciences." In *The Genius of Erasmus Darwin*, edited by C.U.M. Smith and Robert Arnott. Aldershot: Ashgate Publishing.
Tredinnick, Mark. 2005. *The Land's Wild Music: Encounters with Barry Lopez, Peter Matthiessen, Terry Tempest Williams, and James Galvin*. San Antonio: Trinity University Press.
Trigger, David, and Jane Mulcock. 2005. "Forests as Spiritually Significant Places: Nature, Culture and "Belonging" in Australia." *The Australian Journal of Anthropology* 16 (3): 306-320.
Tsing, Anna. 2011. "Arts of Inclusion, or, How to Love a Mushroom." *Australian Humanities Review* 50: 5-21, http://www.australianhumanitiesreview.org/archive/Issue-May-2011/tsing.html.

Tuan, Yi-Fu. 1993. "Desert and Ice: Ambivalent Aesthetics." In *Landscape, Natural Beauty and the Arts*, edited by Salim Kemal and Ivan Gaskell, 139-157. Cambridge: Cambridge University Press.

Turner, Nancy. 1999. "'Time to Burn': Traditional Use of Fire To Enhance Resource Production by Aboriginal Peoples in British Columbia." In *Indians, Fire, and the Land in the Pacific Northwest*, edited by Robert Boyd, 185-218. Corvallis: Oregon State University Press.

Tuteja, Narendra. 2012. *Improving Crop Productivity in Sustainable Agriculture*. Weinheim: Wiley.

UNESCO. 2003. *Convention for the Safeguarding of the Intangible Cultural Heritage*. Paris: UNESCO.

University of Michigan. 2003. "Native American Ethnobotany." Accessed August 19, 2012. http://herb.umd.umich.edu/.

Usher, Peter. 2000. "Traditional Ecological Knowledge in Environmental Assessment and Management." *Arctic* 53 (2): 183-193.

Van den Berg, Rosemary. 2002. *Nyoongar People of Australia: Perspectives on Racism and Multiculturalism*. Leiden: Brill.

Vlamingh, Willem de. 1985. *Voyage to the Great South Land*. Translated by C. De Heer. Sydney: Royal Australian Historical Society.

Walker, J. 1996. "The Classification of the Fungi: History, Current Status and Usage in the *Fungi of Australia*." In *Fungi of Australia*, edited by A. Orchard, 1-27. Canberra: Australian Biological Resources Study; CSIRO.

Western Australia. 2012. *Heritage of Western Australia Act 1990*. Perth: AustLII. http://www.austlii.edu.au/au/legis/wa/consol_act/howaa1990295/.

Wilkes, Ted. 1998. "Interview 25: Ted Wilkes. Perth Aboriginal Medical Service, Director. 12:55pm. 29:06:98." In *Nyoongar Views on Logging Old Growth Forests*, edited by Timothy McCabe, 44-45. West Perth: Wilderness Society.

Willes, John. 1700. *The Julian and Gregorian Year, or, The Difference Betwixt the Old and New-Stile*. London: Richard Sare.

Wolfe, Cary. 2010. *What is Posthumanism?* Minneapolis: University of Minnesota Press.

Wordsworth, William. 2012. "Three Poems on the Celandine by William Wordsworth." Originally published in 1803. Accessed July 31, 2012. http://www.wordsworth.org.uk/poetry/index.asp?pageid=298.

Wright, Judith. 1994. *Judith Wright: Collected Poems 1942-1985*. Sydney: Angus & Robertson.

Wright, Julia. 2012. *Sustainable Agriculture and Food Security in an Era of Oil Scarcity: Lessons from Cuba*. Hoboken: Earthscan.

Yibarbuk, D., P. Whitehead, J. Russell-Smith, D. Jackson, D. Godjuwal, A. Fisher, P. Cooke, D. Choquenot, and D. Bowman. 2001. "Fire Ecology and Aboriginal Land Management in Central Arnhem Land, Northern Australia: A Tradition of Ecosystem Management." *Journal of Biogeography* 28: 325-343.

Index

A

adéquation, 108
Adorno, Theodor 62, 64, 128
aesthesis, 60, 64, 65, 69, *See* aesthetics; embodiment
aesthetics, 19, 22-26, 30, 32, 37, 40, 48, 60, 61-70, 83, 101, 107, 110, 111
allocentric, 64, 65, 69, 92, 95
autocentric, 64

B

balga, 46, 66, 67, 68, 74, 75, 76, *See* grass tree
banksia, 13, 29, 102
Bates, Daisy, 46
Baumgarten, Alexander, 64
Bede, 8, 129, *See* the seasons
being singular plural, 95
being *with*, 1, 3, 38, 49, 59, 60, 63, 70, 86, 91, 100, 106-108, *See* phenomenology
Birok, 12, 13, 14
botanical heritage, 49, 58, 65, *See* plant-based cultural heritage
Bryson, J. Scott, 88, 90, 91, 129, 130
Bureau of Meteorology. *See* the seasons
Burnoru, 12-14

C

Caddy, Caroline, 88, 89, 91, 98, 99, 130
calendar plants, 6
Carlson, Allen, 19, 23-26, 35, 37, 101, 130, *See* aesthetics
catspaw, 56

Choate, Alec. *See* habitat poetry
Colbung, Ken, 45, 130, 132
Collard, Len, 14, *See* the seasons
conservation, 22, 44, 49, 50-54, 58-60, 62, 63, 67, 68, 89, *See* heritage
Convention for the Safeguarding of the Intangible Cultural Heritage, 52, 143
corporeal, 4, 5, 11, 17, 41, 44, 80, 87, 94, 97, 98, 110, 111
corporeality. *See* embodiment
Cosgrove, Denis, 70, 72, 86, 101, 130
critically pluralistic environmental aesthetic, 19, 20, 23, 28, 32, 36, 38

D

Darwin, Erasmus, 114
de Vlamingh, Willem. *See* zamia
Denali, xi, 19-26, 28, 30, 31, 35, 36, 38, 140, *See* aesthetics
Dickinson, Emily, 88, 89, 91, 92, 93, 94, 95, 96, 97, 98, 99, 100, 132
dif-ference, 101-102, 113
Dreaming, 12, 27, 35, 44, 103
Dryandra sessilis, 102
Dukaladjarranj, 28
dwelling, 3-4, 15, 73

E

ecopoetry, 70, 88, 90
embodied temporality, 3, 6, 11, 13, 15, 17, *See* environmental humanities
Entwistle, Tim, 1, 2, 3, 6, 7, 10, 11, 17, 18, *See* the seasons

F

Faith in a Seed, 40, 41, 142, *See* Thoreau, Henry David
FloraCultures, xi, 49-65, 69
fungus, 81, 82, 89-96, 98-100

G

geoautobiography, 110, 111
geoautoethnography, 111
Geran, 12, 13, 14
Giblett, Rod, 70, 71, 72, 132
Gilbert, John, 46
Gregorian. *See* Christian calendar

H

habitat poetry, 71
Haemodoraceae, 56
Haraway, Donna, 88, 91, 97, 98, 100, 133
Heidegger, Martin, 3-6, 18, 72, 101, 102, 105, 108, 113, 132, 134, 136, *See* phenomenology
heritage, 15, 46, 49-63, 69
Hewett, Dorothy. *See* habitat poetry
Heyd, Thomas, 26, 27, 31, 134

I

IKWP, 16, *See* Indigenous Knowledge Weather Project
Indigenous calendar systems, 8, *See* Nyoongar six seasons
Indigenous Weather Knowledge Project, 2, 15, 18, *See* the seasons
intentionality, 6, 95, 96
Inter Gravissimas, 9

J

jarrah, 15, 42, 106, 113
Jilba, 12, 14
Julian drift, 9, 10

K

Kambarang, 12, 14, 15
kangaroo paw, 57, 58, 108
Kant, Immanuel, 60, 63, 64, 102, 135
Kings Park and Botanic Garden, xi, 49, 50, 54, 57
Kinsella, John. *See* habitat poetry
Koyukon, 21, 25, 27
kwongan, 13, 45, 46, 102

L

Landscape, 71, 72, 128-132, 139, 142, 143
Lansdown, Andrew. *See* habitat poetry
Lilio, Aluise Baldassar, 9
Lindley, John, 47, 56, 76, 77, 136

M

Maggoro, 12, 14
materiality, 6, 10, 95, 97, 110, *See* embodiment
Merleau-Ponty, Maurice, 3, 5, 6, 137
Molloy, Georgiana, 57
Moore, George Fletcher. *See* narrative
mudjar, 46, *See* West Australian Christmas Tree
multiple narrative streams, 39, 40, 44, 48
multispecies theory, 88, 90, 91, 94

N

narrative, 19-44, 48, 103, 107, 110, 111, *See* multiple narrative streams
natural environmental model, 19, 24, 25, 26, 37, 101, *See* Carlson, Allen
Nourishing Terrains, 22, 30, 139
Nuytsia, 13, 46, 47, 76-78, 128, 134, *See* West Australian Christmas Tree
Nyoongar, x, 1, 2, 3, 5, 7, 11-18, 40, 42-48, 56, 57, 58, 65-68, 103, 105, 106, 108, 129-138, 143

O

objectification, 5, 63, 64
Oliver, Mary, 91
organic happiness, 122

P

peripateia, 113
Perth, 5, 12, 13, 14, 42, 43, 47-79, 83, 102, 104, 111, 112, 128, 129-143, *See* South-West of Western Australia
phen(omen)ology, 3, 18, *See* phenology; phenomenology
phenological, 1, 3, 5, 16-18, *See* phenology
phenology, 3, 16, 17, 40, 133, *See* phenomenology
phenomenology, 3, 5, 104, 132, 137, 142, *See* being with
Phillips, Glen. *See* habitat poetry
plant-based cultural heritage, 49-58
Plath, Sylvia, 88-99, 129, 134, 139
poesis, 90, 94, 95, 99, 100
poetry, 39, 48, 55, 62, 70-96, 102, 105, 107-117, 122, 124-126, 143
process, 107, 139-141
Punya, 29

R

respect, 50, 54, 91, 97, 100, 105, 106, 133
Rhizanthella gardneri, 81, 82, 129
Rolston, Holmes, 26, 36, 139
Rose, Deborah Bird. *See* nourishing terrain
Ryan, Tracy. *See* habitat poetry

S

Saito, Yuriko, 27, 31, 140
Salvado, Dom Rosendo. *See* the seasons
sandalwood, 83, 84
Schachtel, Ernest. *See* senses
Science, 16, 23, 26, 28, 37, 39, 48, 88, 114, 117, 126, 128, 131, 134, 136-140
Seddon, George, 52, 61, 62, 74, 112, 140
sense, 5, 45, 61, 73, 101, 102, 109, 112, 128, 139, 140, 142, *See* multisensorial experience
senses, 1, x, 39, 48, 62, 88, 90, 141, *See* aesthetics
Sobel, David. *See* aesthetics
South-West, x, 1, 3, 11, 14, 15, 39, 40-48, 56, 61, 62, 70-87, 101, 102, 104, 105, 107, 108, 110, 112, 131, 137, *See* Western Australia
South-West of Western Australia. *See* biodiversity hotspot
sustainability, 50, 54, 58, 60, 61, 65, 68

T

tangible, x, 3, 23, 31, 50, 52, 53, 54, 56, 59
The Botanic Garden, xi, 114-117, 119, 122, 125, 131, *See* Darwin, Erasmus
The Temple of Nature, 114, 116,

122-125, 131, *See* Darwin, Erasmus
Thomas, Suzanne, 110
Thoreau, Henry David, 39, 40, 41, 48, 73, 109, 111, 131, 142

U

Understanding Parrot Bush, 102, 103

W

West Australia Christmas tree, 46, *See* mudja
western North America, 32-33

Wild Fruits, 40-41, 142, *See* Thoreau, Henry David
wildfires, 28, 30
Wright, Judith, xi, 39, 48, 61, 116, 143

Z

zamia, 29, 42, 43, 67, 111, 112, 113, *See* narrative

www.ingramcontent.com/pod-product-compliance
Lightning Source LLC
Chambersburg PA
CBHW070832300426
44111CB00014B/2527